About the Author

Mary Jane Sterling is also the author of *Algebra For Dummies*, *Algebra I CliffStudySolver* and *Algebra II CliffStudySolver*. She taught junior high school and high school math for many years before beginning her current twenty-five-years-and-counting tenure at Bradley University in Peoria, Illinois. Mary Jane especially enjoys working with future teachers and trying out new technology. She keeps busy with Kiwanis and other volunteer activities — when not studying or writing about mathematics. She and her husband, Ted, are enjoying being empty-nesters, although the kids still know where they live.

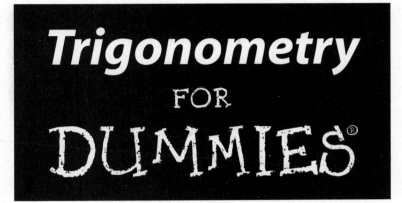

Trigonometry FOR DUMMIES®

by Mary Jane Sterling

WILEY

Wiley Publishing, Inc.

Trigonometry For Dummies®

Published by
Wiley Publishing, Inc.
111 River St.
Hoboken, NJ 07030-5774
www.wiley.com

For general information on our other products and services, please contact our Customer Care Department within the U.S. at 877-762-2974, outside the U.S. at 317-572-3993, or fax 317-572-4002.

For technical support, please visit www.wiley.com/techsupport.

Wiley also publishes its books in a variety of electronic formats. Some content that appears in print may not be available in electronic books.

Library of Congress Control Number: 2004116322

ISBN: 978-0-7645-6903-6

Manufactured in the United States of America

15 14 13 12

WILEY

Dedication

The author would like to dedicate *Trigonometry For Dummies* to all the men and women serving in the United States armed services, both at home and abroad. The appreciation we have for their dedication and sacrifice can never be adequately measured.

Publisher's Acknowledgments

We're proud of this book; please send us your comments through our Dummies online registration form located at www.dummies.com/register/.

Some of the people who helped bring this book to market include the following:

Acquisitions, Editorial, and Media Development

Senior Project Editor: Tim Gallan

Acquisitions Editor: Kathy Cox

Copy Editor: Kristin DeMint

Editorial Program Assistant: Courtney Allen

Technical Editor: David Herzog

Editorial Manager: Christine Meloy Beck

Editorial Assistants: Hanna Scott, Melissa S. Bennett, Nadine Bell

Cover Photo: © Getty Images/Photodisc Blue

Cartoons: Rich Tennant, www.the5thwave.com

Composition

Project Coordinator: Adrienne Martinez

Layout and Graphics: Andrea Dahl, Lauren Goddard, Denny Hager, Joyce Haughey, Jacque Roth, Mary Gillot Virgin

Proofreaders: Leeann Harney, Carl Pierce, Kathy Simpson, Charles Spencer

Indexer: Steve Rath

Publishing and Editorial for Consumer Dummies

    Diane Graves Steele, Vice President and Publisher, Consumer Dummies

    Joyce Pepple, Acquisitions Director, Consumer Dummies

    Kristin A. Cocks, Product Development Director, Consumer Dummies

    Michael Spring, Vice President and Publisher, Travel

    Brice Gosnell, Associate Publisher, Travel

    Kelly Regan, Editorial Director, Travel

Publishing for Technology Dummies

    Andy Cummings, Vice President and Publisher, Dummies Technology/General User

Composition Services

    Gerry Fahey, Vice President of Production Services

    Debbie Stailey, Director of Composition Services

Contents at a Glance

Table of Contents

Introduction

Some of the more-practical accomplishments of early humankind were made possible by trigonometry. Even before mathematicians formalized trigonometry into a particular area of study or used it to solve problems, trigonometry helped people sail across large bodies of water, build structures, plot out land, and measure heights and distances.

People still use trigonometry for all these reasons and more. If you're going to get your pilot's license, you need trigonometry. This subject is also the basis for many courses in mathematics, including calculus. Trig is all over the place because it's a very versatile subject. It brings in the best from the worlds of algebra and geometry and makes for some interesting reading and doing.

You can get as deeply into this book as you want — you can even wade through it — and you'll still come out of it thinking, "Gee. I didn't realize that was trigonometry! Wasn't that just loads of fun!" Well, *loads* may be a bit of an exaggeration.

Whether you're pursuing trigonometry so that you can go on to study calculus, prepare for architecture or drafting, or become a world-traversing pilot, you'll find all that you need to know in *Trigonometry For Dummies*. You can get as technical as you want, skipping over the stuff you don't need. Welcome to the mathematical adventure that humankind embarked on many years ago — without even knowing it!

About This Book

You may be wondering what in the world a book on trigonometry will have to ring your bell, tickle your fancy, or just make you pretty happy. Oh, where to begin?

If you're really excited about triangles and how you can use them to your advantage, this book is for you. What if you're looking for angles? Dear reader, you've come to the right place. I show you how to find angles of all shapes and sizes: big ones and little ones, all named depending on their situation, and measured in degrees or radians. "What's a radian?" you ask. You can find out in this book, that's for sure. If you are analytically minded, you'll find plenty to discover about identities and equation solving. And if art is your bag, this book has simple trig graphs, complicated trig graphs, and everything in between.

Conventions Used in This Book

To help you navigate through this book, I've set up a few conventions.

- ✔ Because the words in math are very precise, I point them out with *italics* so you realize that you've stumbled onto an important term or idea. If you simply want to look up a term and don't want to search through the chapters for a definition, flip on over to the glossary.
- ✔ I use all the usual math symbols for addition, subtraction, multiplication, division, and roots. All the symbols are pretty standard — nothing too jazzy.
- ✔ I provide lots and lots of examples, formatted step-by-step with explanations along the way.

Foolish Assumptions

What kind of fool am I to assume that you're reading *Trigonometry For Dummies* because it looks more interesting than the latest bestseller? I'm not that foolish! To be honest, trig wouldn't be my first choice for a fun read. I'm just going to assume that you really want to do this. While writing this book, I made a few other assumptions about you as well:

- ✔ You have a goal in mind: You want to conquer some of the topics in this book so you're prepared for a course of study that you have in mind.
- ✔ You have a pretty solid grasp on algebra and can solve a simple algebraic equation without falling completely apart.
- ✔ You're planning on being on a quiz show such as *Who Wants to Be a Millionaire?* or *Jeopardy!* and need to bone up on the possible trig questions.

What You're Not to Read

Unlike literature books, you *don't* have to read this book from cover to cover. The chapters pretty much stand by themselves, so you don't have to do a lot of review. I try to explain the concepts as clearly as possible, putting stuff into down-to-earth terms so that it's fun *and* instructive. By the way, the text in gray boxes (also known as sidebars in the *Dummies* universe) is completely optional reading. Sometimes I go off on a tangent, and sidebars are good for that sort of thing.

How This Book Is Organized

This book really *is* organized. The different topics are grouped together, and they follow a logical flow.

Part 1: The Basics

This part has all the geometry, especially triangles, that you need if you want to review. You'll find the different shapes, forms, and constructions of triangles that are most helpful in trigonometry — but the rest of geometry, I leave out. I show you the many ways to measure an angle and, as a refresher, zone in on the big difference between a degree and a radian. Because trig is all about functions, I provide a primer on functions in general. And as you probably expected, this book has lots of equations and graphs, complete with discussion so you're not left in the dark.

Part II: Trigonometric Functions

This part introduces the ways that the six trig functions came to be — first, I show you by using right triangles and then, the unit circle. The right triangles are easier to grasp, and Pythagoras (remember him from algebra?) is involved. The unit circle is necessary because you want to be able to work outside the box — oops, triangle. The right-triangle definitions lend themselves to some nice applications, which are also in this part.

Part III: Identities

Here, you get to the heart of trigonometry as you discover how the different functions are interrelated. I introduce the identities in groupings that help you remember them and figure out when to apply them. And for you curious types who simply *must* know where everything came from, I include a whole chapter on solving these identities to show how they came to be and how you can use them.

Part IV: Equations and Applications

This part starts out with the inverse trig functions, which are quite handy when you're solving equations with trig functions in them. You discover many different techniques for solving these different types of trig equations, and I give lots of examples to get you started on the right foot. Then I introduce the law of sines and the law of cosines, which flow naturally into different ways to find the area of a triangle.

Part V: Graphs of Trig Functions

The title of this part tells all. You find each of the graphs of the six basic functions, along with the graphs of the six inverse trig functions. To round out these chapters, I show you how to perform different variations on these graphs.

Part VI: The Part of Tens

Four lists in this part provide, in a quick-reference format, all the identities and equations you need to solve trig problems (and then some). Flip to these lists of 10 (one is actually a list of 11) to find the basic trig identities, some obscure identities, ways to determine the area of a triangle — especially when it's really big or inconvenient to measure with a ruler — and, lastly, some neat trig equations you can use to predict weather, with at least the accuracy of that weatherperson on TV.

Icons Used in This Book

The icons, which point out particular kinds of information that may be of use to you, are easy to spot. The ones you find in this book are as follows:

Remember icons either give you a heads-up on an important trig concept or refer to information that I think you may already know from previous math courses. I point out those concepts so that the new stuff makes sense.

A tip is just what you think: a bit of advice for understanding a concept or solving a problem.

Of course, trig rules! Aside from that reminder, this icon points out particular equations, expressions, formulas, identities — what have you — that you should pay attention to. They're important.

In any area of mathematics, you always come across concepts that are tricky or confusing or problems that practically beg you to make an error. This icon is here to alert you, helping you avoid a mathematical pitfall.

Where to Go from Here

Back when I was in college, my friend Judy Christopher once consoled me with the saying, "Life is like a sine curve; it has its ups and downs. If you're feeling really down, then just remember that you'll be going up that same amount someday soon." So maybe you want to start with the graphs of the sine curves and other trig curves. Make of them what you will.

Or maybe, like me, you're a puzzle buff. Nothing gives you more sheer joy than gathering bits and pieces of information and making them all fit into a logical pattern. If that idea strikes your fancy, then go to the chapters on proving identities and solving equations. This book provides a great challenge for a rainy Sunday afternoon's pleasure.

Are you into angles, directions, and plans? You may want to start with the many ways you can measure angles and how they all fit together in the big picture of everyday life. The basics are always a good place to start when you're investigating a topic.

No matter where you start in this book, be ready to flip the pages front to back or back to front. Think of it as an adventure that can take you to many interesting places. Enjoy!

Part I
The Basics

In this part . . .

Trigonometry is closely tied to geometry — with a smattering of algebra thrown in. These chapters get you started with the geometry of triangles and the basics of measuring angles in both degrees and radians. Top it off with some graphs and formulas, and you're good to go.

Chapter 1

Troucing Trig Technicalities

How did Columbus find his way across the Atlantic Ocean? How did the Egyptians build the pyramids? How did early astronomers measure the distance to the moon? No, Columbus didn't follow a yellow brick road. No, the Egyptians didn't have Legos instructions. And, no, there isn't a tape measure long enough to get to the moon. The common answer here is trigonometry.

Trigonometry is the study of angles and triangles and the wonderful things that you can do with them. For centuries, humans have been able to measure distances that they can't reach because of the power of this mathematical subject.

Taking Trig for a Ride: What Trig Is

"What's your angle?" That question isn't a come-on such as "What's your astrological sign?" In trigonometry, you measure angles in both degrees and radians. You can shove them into triangles and circles and make them do special things. Actually, angles drive trigonometry. Sure, you have to consider algebra and other math. But you can't have trigonometry without angles. Put an angle into a trig function, and out pops a special, unique number. What do you do with that number? Read on, because that's what trig is all about.

Sizing up the basic figures

Segments, rays, and lines are some of the basic forms in geometry, and they're almost as important in trigonometry. As I explain in the following sections, you use those segments, rays, and lines to form angles.

Drawing segments, rays, and lines

A *segment* is a straight figure drawn between two endpoints. You usually name it by its endpoints, which you indicate by capital letters. Sometimes, a single letter names a segment, but a lowercase letter usually refers to an angle opposite that segment.

A *ray* is another straight figure that has an endpoint on one end, and then it just keeps going forever in some specified direction. You name rays by their endpoint first and then by any other point that lies on the ray.

A *line* is a straight figure that goes forever and ever in either direction. You only need two points to determine a particular line — and only one line can go through both of those points. You can name a line by any two points that lie on it.

Figure 1-1 shows a segment, ray, and line.

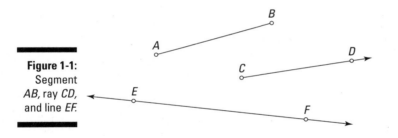

Figure 1-1:
Segment
AB, ray *CD,*
and line *EF.*

Intersecting lines

When two lines intersect — if they do intersect — they can only do so at one point. They can't double back and cross one another again. Some curious things happen when two lines intersect. The angles that form between those two lines are related to one another. Any two angles that are next to one another and share a side are called *adjacent* angles. In Figure 1-2, the lines *AB* and *CD* intersect at point *E*. The two angles below the line *CD* (numbered 1 and 2) are adjacent to one another. So are the two angles to the right of line *AB* (numbered 2 and 3), the angles to the left of line *AB* (numbered 1 and 4), and the angles above line *CD* (numbered 3 and 4). So this intersection has four different pairs of adjacent angles.

The angles that are opposite one another when two lines intersect also have a special name. Mathematicians call these angles *vertical* angles. They don't have a side in common. You can find two pairs of vertical angles in Figure 1-2, the pair of angles to the left and right (numbered 1 and 3), and the pair above and below (numbered 2 and 4).

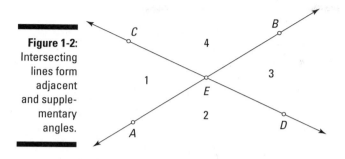

Figure 1-2: Intersecting lines form adjacent and supplementary angles.

Why are these different angles so special? They're different because of how they interact with one another. The adjacent angles here are called *supplementary* angles. The sides that they don't share form a straight line, which has a measure of 180 degrees. The vertical angles are always equal in measure.

Angling for position

When two lines, segments, or rays touch or cross one another, they form an angle. In the case of two intersecting lines, the result is four different angles. When two segments intersect, they can form one, two, or four angles, depending on how they touch, as you can see in Figure 1-3. The same goes for two rays.

These examples are just some of the ways that you can form angles. Geometry, for example, describes an angle as being created when two rays have a common endpoint. In practical terms, you can form an angle in many ways, from many figures. The business with the two rays means that you can extend the two sides of an angle out farther to help with measurements, calculations, and practical problems.

You refer to the parts of all angles in the same way. The place where the lines, segments, or rays cross is called the *vertex* of the angle. From the vertex, two sides extend.

Naming angles by size

You can name or categorize angles based on their size or measurement in degrees (see Figure 1-4):

- ✔ **Acute:** An angle measuring between 0 and 90 degrees
- ✔ **Obtuse:** An angle measuring between 90 and 180 degrees
- ✔ **Right:** An angle measuring exactly 90 degrees
- ✔ **Straight:** An angle measuring exactly 180 degrees (a straight line)

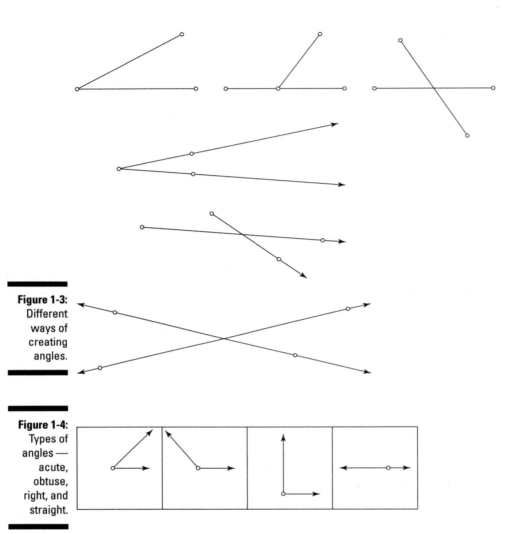

Figure 1-3: Different ways of creating angles.

Figure 1-4: Types of angles — acute, obtuse, right, and straight.

Naming angles by letters

How do you name an angle? Why does it even need a name? In most cases, you want to be able to distinguish a particular angle from all the others in a picture. When you look at a photo in a newspaper, you want to know the names of the different people and be able to point them out. With angles, you should feel the same way.

You can name an angle in one of three different ways:

✔ **By its vertex alone:** Often, you name an angle by its vertex alone because such a label is efficient, neat, and easy to read. In Figure 1-5, you can call the angle *A*.

✔ **By a point on one side, followed by the vertex, and then a point on the other side:** For example, you can call the angle in Figure 1-5 angle *BAC* or angle *CAB*. This naming method is helpful if someone may be confused as to which angle you're referring to in a picture. **Remember:** Make sure you always name the vertex in the middle.

✔ **By a letter or number written inside the angle:** Usually, that letter is Greek; in Figure 1-5, however, the angle has the letter *w*. Often, you use a number for simplicity if you're not into Greek letters or if you're going to compare different angles later.

Figure 1-5:
Naming an
angle.

Triangulating your position

All on their own, angles are certainly very exciting. But put them into a triangle, and you've got icing on the cake. Triangles are one of the most frequently studied geometric figures. The angles that make up the triangle give them many of their characteristics.

Angles in triangles

A triangle always has three angles. The angles in a triangle have measures that always add up to 180 degrees — no more, no less. A triangle named *ABC* has angles *A*, *B*, and *C*, and you can name the sides *AB*, *BC*, and *AC*, depending on which two angles the side is between. The angles themselves can be acute, obtuse, or right. If the triangle has either an obtuse or right angle, then the other two angles have to be acute.

Naming triangles by their shape

Triangles have special names based on their angles and sides. They can also have more than one name — a triangle can be both acute and isosceles, for example. Here are their descriptions, and check out Figure 1-6 for the pictures:

- ✔ **Acute triangle:** A triangle where all three angles are acute

- ✔ **Right triangle:** A triangle with a right angle (the other two must be acute)

- ✔ **Obtuse triangle:** A triangle with an obtuse angle (the other two must be acute)

- ✔ **Isosceles triangle:** A triangle with two angles that have the same measure; the lengths of the sides opposite those angles are equal, too

- ✔ **Equilateral triangle:** A triangle where all three angles measure 60 degrees; all the lengths of the sides are equal, too

- ✔ **Scalene triangle:** A triangle with no angles or sides of the same measure

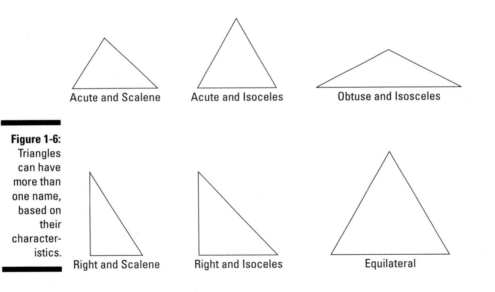

Acute and Scalene Acute and Isoceles Obtuse and Isosceles

Figure 1-6: Triangles can have more than one name, based on their characteristics.

Right and Scalene Right and Isoceles Equilateral

Circling the wagons

A *circle* is a geometric figure that needs only two parts to identify it and classify it by size: its *center* (or middle) and its *radius* (the distance from the center to any point on the circle). Technically, the center isn't a part of the circle, it's just a sort of anchor or reference point. The circle consists only of all those points that are the same distance from the center.

Radius, diameter, circumference, and area

After you've chosen a point to be the center of a circle and know how far that point is from all the points that lie on the circle, you can draw a fairly decent picture. With the measure of the radius, you can tell a lot about the circle: its *diameter* (the distance from one side to the other, passing through the center), its *circumference* (how far around it is), and its *area* (how many square inches, feet, yards, meters — what have you — fit into it). Figure 1-7 shows these features.

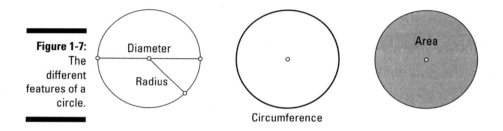

Figure 1-7: The different features of a circle.

Ancient mathematicians figured out that the circumference of a circle is always a little more than three times the diameter of a circle. Since then, they narrowed that "little more than three times" to a value called *pi* (pronounced "pie"), designated by the Greek letter π. The decimal value of *pi* isn't exact — it goes on forever and ever, but most of the time, people refer to it as being approximately 3.14 or $\frac{22}{7}$, whichever form works best in specific computations.

The formula for figuring out the circumference of a circle is tied to π and the diameter.

Circumference of a circle: $C = \pi d = 2\pi r$

The *d* represents the measure of the diameter, and *r* represents the measure of the radius. The diameter is always twice the radius, so either form of the equation works.

Similarly, the formula for the area of a circle is tied to π and the radius.

Area of a circle: $A = \pi r^2$

This formula reads, "Area equals *pi* are squared." And all this time, I thought that pies are round.

Don't give me that *jiva*

The ancient Greek mathematician Ptolemy was born some time at the end of the first century. Ptolemy based his version of trigonometry on the relationships between the chords of circles and the corresponding central angles of those chords. Ptolemy came up with a theorem involving four-sided figures that you can construct with the chords. In the meantime, mathematicians in India decided to use the measure of *half* a chord and *half* the angle to try to figure out these relationships. Drawing a radius from the center of a circle through the middle of a chord (halving it) forms a right angle, which is important in the definitions of the trig functions. These half-measures were the beginning of the sine function in trigonometry. In fact, the word *sine* actually comes from the Hindu name *jiva*.

Example: Find the radius, circumference, and area of a circle if its diameter is equal to 10 feet in length.

If the diameter *(d)* is equal to 10, you write this value as *d = 10*. The radius is half the diameter, so the radius is 5 feet, or *r = 5*. You can find the circumference by using the formula $C = \pi d = \pi \cdot 10 \approx 3.14 \cdot 10 = 31.4$. So the circumference is about 31½ feet around. You find the area by using the formula $A = \pi r^2 = \pi \cdot 5^2 = \pi \cdot 25 \approx 3.14 \cdot 25 = 78.50$, so the area is about 78½ square feet.

Chord versus tangent

You show the diameter and radius of a circle by drawing segments from a point on the circle either to or through the center of the circle. But two other straight figures have a place on a circle. One of these figures is called a chord, and the other figure is a tangent.

Chords of a circle

A *chord* of a circle is a segment that you draw from one point on the circle to another point on the circle (see Figure 1-8). This segment always stays inside the circle. The largest chord possible is the diameter — you can't get any longer than that segment.

Tangents to a circle

A *tangent* to a circle is a line, ray, or segment that touches the outside of the circle in exactly one point, as in Figure 1-9. It never crosses into the circle. A tangent can't be a chord, because a chord touches a circle in two points, crossing through the inside of the circle.

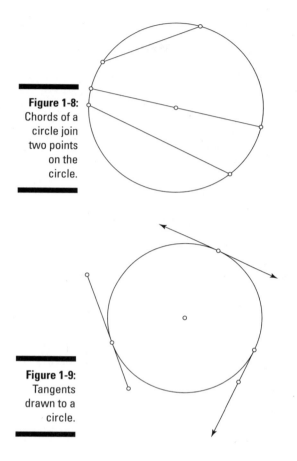

Figure 1-8:
Chords of a
circle join
two points
on the
circle.

Figure 1-9:
Tangents
drawn to a
circle.

Sectioning sectors

A *sector* of a circle is a section of the circle between two *radii* (plural for radius). You can consider this part like a piece of pie cut from a circular pie plate (see Figure 1-10).

You can find the area of a sector of a circle if you know the angle between the two radii. A circle has a total of 360 degrees all the way around the center, so if a sector has an angle measure of 60 degrees between the two radii, the sector takes up 60/360, or ⅙, of the degrees all the way around. In that case, the sector has ⅙ the area of the whole circle.

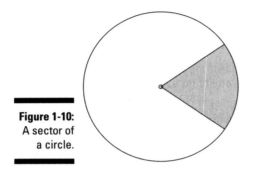

Example: Find the area of a sector of a circle if the angle between the two radii forming the sector is 80 degrees and the diameter of the circle is 4.5 inches.

1. **Find the area of the circle.**

   The area of the whole circle is $A = \pi r^2 = \pi \cdot (4.5)^2 \approx 3.14 \cdot 20.25 = 63.585$, or about 63½ square inches.

2. **Find the portion of the circle that the sector represents.**

   The sector takes up only 80 degrees of the circle. Divide 80 by 360 to get

   $$\frac{80}{360} = \frac{2}{9} \approx 0.222$$

3. **Calculate the area of the sector.**

   Multiply the fraction or decimal from Step 2 by the total area to get the area of the sector: $0.222 \cdot 63.585 \approx 14.116$. The whole circle has an area of almost 64 square inches, but the sector has an area of just over 14 square inches.

Understanding Trig Speak

Any math or science topic has its own unique vocabulary. Some very nice everyday words have special meanings when used in the context of that subject. Trigonometry is no exception.

Defining trig functions

Every triangle has six parts: three sides and three angles. If you measure the sides and then pair up those measurements (taking two at a time), you have three different pairings. Do division problems with the pairings — changing the order in each pair — and you have six different answers. These six different answers represent the six trig functions. For example, if your triangle has sides measuring 3, 4, and 5, then the six divisions are 3/4, 4/3, 3/5, 5/3, 4/5, and 5/4. In Chapter 7, you find out how all these fractions work in the world of trig functions by using the different sides of a right triangle. And then, in Chapter 8, you take a whole different approach as you discover how to define the trig functions with a circle.

The six trig functions are named *sine, cosine, tangent, cotangent, secant,* and *cosecant.* Many people confuse the spoken word *sine* with *sign* — you can't really tell the difference when you hear it unless you're careful with the context. You can "go off on a tangent" in some personal dealings, but that phrase has a whole different meaning in trig. Cosigning a loan isn't what trig has in mind, either. The other three ratios are special to trig speak. You can't confuse them with anything else.

Interpreting trig abbreviations

Even though the word *sine* isn't all that long, you have a three-letter abbreviation for this trig function and all the others. Mathematicians find using abbreviations easier, and those versions fit better on calculator keys. The functions and their abbreviations are

- ✔ sine → sin
- ✔ cosine → cos
- ✔ tangent → tan
- ✔ cotangent → cot
- ✔ secant → sec
- ✔ cosecant → csc

As you can see, the first three letters in the full name make up the abbreviations, except for cosecant's.

Noting notation

Angles are the main focus in trigonometry, and you often don't know their measure. Many angles and their angle measures have general rules that apply to them. You can name angles by one letter, three letters, or a number, but to do trig problems and computations, mathematicians commonly refer to the angle measures with Greek letters.

The most commonly used letters for angle measures are α (alpha), β (beta), γ (gamma), and θ (theta). Also, many equations use the variable x to represent an angle measure.

Algebra has conventional notation involving superscripts, such as the 2 in x^2. In trigonometry, superscripts have the same rules and characteristics as in other mathematics. But trig superscripts often look very different. Table 1-1 presents a listing of many of the ways that trig uses superscripts.

Table 1-1	How You Use Superscripts in Trig
How to Write in Trig Notation	*What the Superscript Means*
$\sin^2 \theta$	$(\sin \theta)^2$
$\sin^{-1} \theta$	$\arcsin \theta$

The first entry in Table 1-1 shows how you can save having to write parentheses every time you want to raise a trig function to a power. This notation is neat and efficient, but it can be confusing if you don't know the "code." The second entry shows you how to write the reciprocal of a trig function. It means you should take the value of the function and divide it into the number 1. The last entry in Table 1-1 shows how you write the *inverse sine* function. Using the –1 superscript right after sine means that you're talking about inverse sine (or *arcsin*), not the reciprocal of the function. In Chapter 13, I cover the inverse trig functions in great detail, making this business about the notation for an inverse trig function even more clear.

Functioning with angles

The functions in algebra use many operations and symbols that are different from the add, subtract, multiply, and divide signs in arithmetic. For example, take a look at the square-root operation, $\sqrt{25} = 5$. Putting 25 under the *radical* (square-root symbol) produces an answer of 5. Other operations in algebra, such as absolute value, factorial, and step-function, are used in trigonometry, too. But the world of trig expands the horizon, introducing even more different processes. When working with trig functions, you have a whole new set of values to learn or find. For instance, putting 25 into the sine function looks like this: sin 25. The answer that pops out is either 0.423 or –0.132, depending on whether you're using degrees or radians (for more on those two important trig concepts, head on over to Chapters 4 and 5). You can't usually determine or memorize the values that you get by putting angle measures into trig functions very easily. So you need trig tables of values or scientific calculators to study trigonometry.

In general, when you apply a trig function to an angle measure, you get some real number. Some angles and trig functions have nice values, but most don't. Table 1-2 shows the trig functions for a 30-degree angle.

Table 1-2	The Trig Functions for a 30-Degree Angle
Trig Function	*Value Rounded to Three Decimal Places*
sin 30°	0.500
cos 30°	0.866
tan 30°	0.577
cot 30°	1.732
sec 30°	1.155
csc 30°	2.000

Some characteristics that the entries in Table 1-2 confirm are that the sine and cosine functions always have values that are between and including –1 and 1. Also, the secant and cosecant functions always have values that are equal to or greater than 1 or equal to or less than –1. I discuss these properties in more detail in Chapter 7.

Using the table in the Appendix, you can find more values of trig functions for particular angle measures (in degrees):

tan 45° = 1

csc 90° = 1

sec 60° = 2

I chose these sample values so the answers look nice and whole. Remember that most angles and most functions look much messier than these examples.

Taming the radicals

A *radical* is a mathematical symbol that means, "Find the number that multiplies itself by itself one or more times to give you the number under the radical." You can see why you use a symbol such as $\sqrt{}$ rather than all those words. Radicals represent functions that are used a lot in trigonometry. In Chapter 7, I define the trig functions by using a right triangle. To solve for the lengths of a right triangle's sides by using the Pythagorean theorem, you have to compute some square roots, which use radicals. Some basic answers to radical expressions are $\sqrt{16} = 4$, $\sqrt{121} = 11$, $\sqrt[3]{8} = 2$, $\sqrt[4]{6561} = 9$.

These examples are all *perfect squares, perfect cubes,* or *perfect fourth roots,* which means that the answer is a number that ends — the decimal doesn't go on forever. The following section discusses a way to simplify radicals that aren't perfect roots.

Simplifying radical forms

Simplifying a radical form means to rewrite it with a smaller number under the radical — if possible. You can simplify this form only if the number under the radical has a perfect square or perfect cube (or perfect whatever factor) that you can factor out.

Example: Simplify $\sqrt{80}$.

The number 80 isn't a perfect square, but one of its factors, 16, is. You can write the number 80 as the product of 16 and 5, write the two radicals separately, and then evaluate each radical. The resulting product is the simplified form:

$$\sqrt{80} = \sqrt{16 \cdot 5} = \sqrt{16}\sqrt{5} = 4\sqrt{5}$$

Example: Simplify $\sqrt[3]{250}$.

The number 250 isn't a perfect cube, but one of its factors, 125, is. Write 250 as the product of 125 and 2; separate, evaluate, and write the simplified product:

$$\sqrt[3]{250} = \sqrt[3]{125 \cdot 2} = \sqrt[3]{125}\sqrt[3]{2} = 5\sqrt[3]{2}$$

Approximating answers

As wonderful as a simplified radical is, and as useful as it is when you're doing further computations in math, sometimes you just need to know about how much the value's worth.

Approximating an answer means to shorten the actual value in terms of the number of decimal places. You may find approximating especially useful when the decimal value of a number goes on forever without ending or repeating. When you approximate an answer, you *round* it to a certain number of decimal places, letting the rest of the decimal values drop off. Before doing that, though, you need to consider how big a value you're dropping off. If the numbers that you're dropping off start with a 5 or greater, then bump up the last digit that you leave on by 1. If what you're dropping off begins with a 4 or less, then just leave the last remaining digit alone.

They called this simpler?

Some ancient mathematicians didn't like to write fractions unless they had a numerator of 1. They only liked the fractions ½, ⅓, ¼, ⅕, and so on. So what did they do when they needed to write the fraction ⅚? They wrote ½ + ⅓ instead (because ½ + ⅓ is equal to ⅚). What a pain to have to write ½ + ¼ + ⅒ rather than ¹⁷⁄₂₀. Or maybe you prefer this approach, too?

Example: Round the number 3.141592654 to four decimal places, three decimal places, and two decimal places.

✔ **Four decimal places:** This rounding value means that the 3.1415 stays. Because you get to drop off the 92654, and 9 is the first digit of those dropped numbers, bump up the last digit that you're keeping (the 5) to 6. Rounded to four places, 3.141592654 rounds to 3.1416.

✔ **Three decimal places:** The 3.141 stays. Because you drop off the 592654, and 5 is the first digit of those numbers, bump up the last digit that you're keeping (the 1) to 2. Rounded to three places, 3.141592654 rounds to 3.142.

✔ **Two decimal places:** The 3.14 stays. Because the 1592654 drops off, and 1 is the first digit of those numbers, then the last digit that you're keeping (the 4) stays the same. Rounded to two places, 3.141592654 rounds to 3.14.

Use this technique when approximating radical values. Using a calculator, the decimal value of $\sqrt{80}$ is about 8.94427191. Depending on what you're using this value for, you may want to round it to two, three, four, or more decimal places. Rounded to three decimal places, this number is 8.944.

Equating and Identifying

Trigonometry has the answers to so many questions in engineering, navigation, and science. The ancient astronomers, engineers, farmers, and sailors didn't have the current systems of symbolic algebra and trigonometry to solve their problems, but they did well and set the scene for later mathematical developments. People today benefit big-time by having ways to solve equations in trigonometry that are quick and efficient, including special techniques and trig identities to fool around with that mathematicians in days of old have already worked out.

The methods that you use for solving equations in algebra take a completely different turn when you use trig *identities* (in short, equivalences that you can substitute into equations in order to simplify them). To make matters easier (or, some say, to *complicate* them), the different trig functions can have many different identities. They almost have split personalities. When you're solving trig equations and trig identities, you're sort of like a detective working your way through to substitute, simplify, and solve. What answer should you expect when solving the equations? Why, angles, of course!

Take, for example, one trigonometric equation: $\sin\theta + \cos^2\theta = 1$.

The point of the problem is to figure out what angle or angles should replace the θ to make the equation true. In this case, if θ is 0 degrees, 90 degrees, or 180 degrees, for example, the equation is true.

If you replace θ by 0 degrees in the equation, then you get

$$\sin 0° + \left(\cos 0°\right)^2 = 1$$
$$0 + \left(1\right)^2 = 1$$
$$1 = 1$$

If you replace θ by 90 degrees in the equation, then you get

$$\sin 90° + \left(\cos 90°\right)^2 = 1$$
$$1 + \left(0\right)^2 = 1$$
$$1 = 1$$

Something similar happens with 180 degrees and all the other angle measures that work in this equation. But remember that not just any angle will work here. I carefully chose the angles that are *solutions,* which are the angles that make the equation true. In order to solve trig equations like this one, you have to use inverse trig functions, trig identities, and algebra techniques. You can find all the details on how to use these processes in Chapters 10 through 13. And when you've got those parts figured out, dive into Chapter 14, where the equation-solving comes in.

In this particular case, you use an *identity* to solve the equation for all its answers. You replace the $\cos^2\theta$ by $1 - \sin^2\theta$ so that all the terms have a sine in them — or just a number. You actually have other ways to change the identity of $\cos^2\theta$, too. I chose $1 - \sin^2\theta$, but other choices include $\dfrac{1}{\sec^2\theta}$ and $\dfrac{1+\cos 2\theta}{2}$. Discover how to actually solve equations like this one in Chapter 14.

This example just shows you that the identity of the trig functions can change an expression significantly — according to some very strict rules.

Graphing for Gold

The trig functions have distinctive graphs that you can use to help understand their values over certain intervals and in particular applications. In this section, I describe the axes and show you six basic graphs.

Describing graphing scales

You use the *coordinate plane* for graphing in algebra, geometry, and other mathematical topics. The *x*-axis goes left and right, and the *y*-axis goes up and down. You can also use the coordinate plane in trigonometry, with a little added twist.

The *x*-axis in a trig sketch has tick marks that can represent both numbers (either positive or negative) and angle measures (either in degrees or radians). You usually want the horizontal and vertical tick marks to have the same distance between them. To make equivalent marks on the *x*-axis in degrees, figure that every 90 degrees is about 1.6 units (the same units that you're using on the vertical axis). These *units* represent numbers in the real-number system. This conversion method works because of the relationship between degree measure and radian measure. Check out the method used to do the computation for this conversion in Chapter 5.

Recognizing basic graphs

The graphs of the trig functions have many similarities and many differences. The graphs of the sine and cosine look very much alike, as do the tangent and cotangent, and then the secant and cosecant. But those three groupings look different from one another. The one characteristic that ties them all together is the fact that they're *periodic,* meaning they repeat the same curve or pattern over and over again, in either direction along the *x*-axis. Check out Figures 1-11 through 1-16 to see for yourself.

I say a lot more in this book about the trig-function graphs, and you can find that discussion in Chapters 16, 17, 18, and 19.

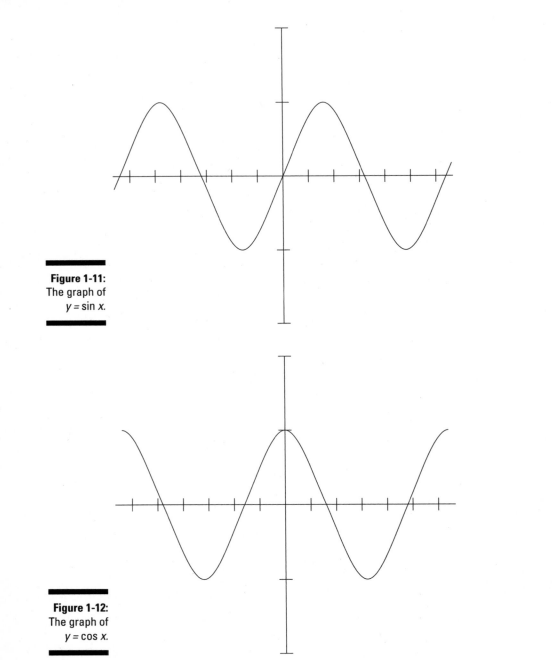

Figure 1-11:
The graph of
$y = \sin x$.

Figure 1-12:
The graph of
$y = \cos x$.

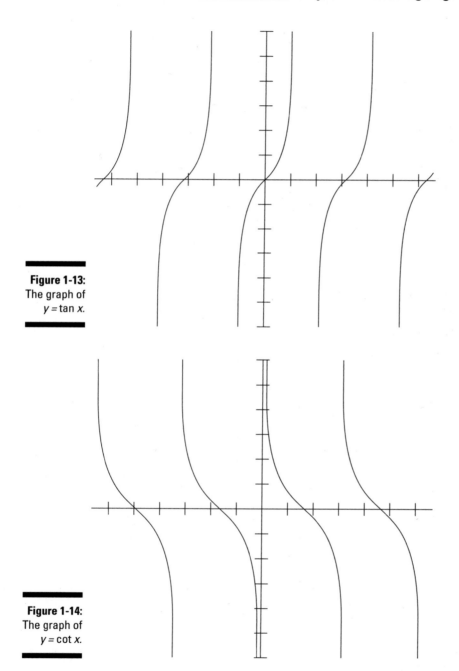

Figure 1-13:
The graph of
$y = \tan x.$

Figure 1-14:
The graph of
$y = \cot x.$

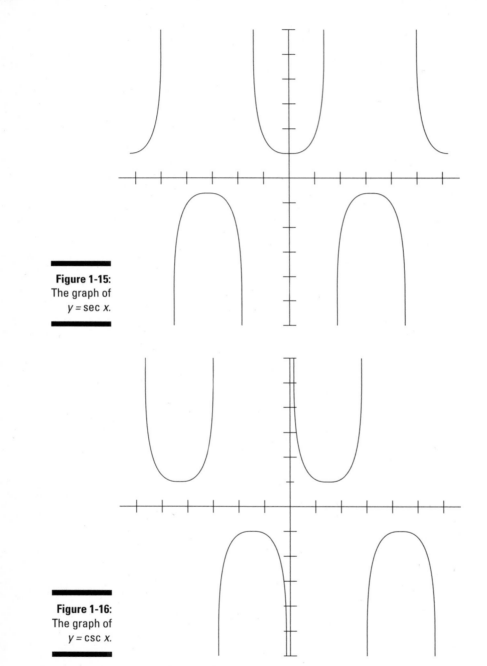

Figure 1-15:
The graph of
$y = \sec x$.

Figure 1-16:
The graph of
$y = \csc x$.

Chapter 2

Coordinating Your Efforts with Cartesian Coordinates

A picture is worth a thousand words. Drawing pictures or graphs of functions and equations in math helps you understand what's going on with them. In trigonometry, you often draw angles and triangles, in addition to the curves that represent the trig functions (sine, cosine, tangent, cotangent, secant, and cosecant). The standard Cartesian coordinate system, which you use when drawing graphs in algebra and other math topics, works best here. If you're looking for a refresher on this point-plotting system, you can find it in this chapter. In short, everything reads from left to right and from bottom to top, running through the negative to the positive numbers.

Starting Out Simple: Plotting Points

Plotting points on a mathematical graph means finding the correct position for a dot that represents an ordered pair of numbers, such as (2,3), (−1,4), or (0,0). This ordered pair is called the *Cartesian coordinates* of the point. You start with two intersecting lines called *axes*.

Axes, axes, we all fall down

Plotting points and drawing graphs requires two axes and a defined distance or scale on them. The two intersecting, perpendicular lines that make up a graph are called the horizontal and vertical *axes* (or *coordinate axes*). These lines extend left and right, up and down, without end. The horizontal axis is traditionally known as the *x*-axis, although in trigonometry the horizontal axis is often labeled the θ axis. The vertical axis is the *y*-axis. Figure 2-1 shows the two axes intersecting at the *origin,* labeled *O.* The part of the *x*-axis going to the right stands for positive numerical values, and you use it as the starting place or initial side when drawing angles in the *standard position.*

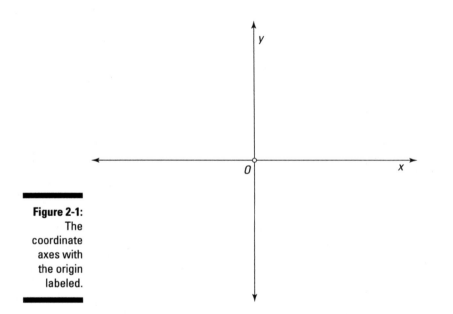

Figure 2-1:
The coordinate axes with the origin labeled.

Determining the origin of it all

The point where the two axes cross is called the *origin.* You label it with an *O* or with its ordered pair of numbers, also called its Cartesian coordinates, (0,0). The origin is the starting point for counting off the distances when graphing all other points. It's also the endpoint of the *rays* (lines that extend infinitely in one direction) that you use when drawing angles in the standard position on the coordinate axes.

Putting da cart before da horse

Rene Descartes was considered to be a mover and shaker in the 17th-century scientific community. He was responsible for many innovations in algebra and geometry. He's also credited with creating our coordinate system for graphing in these mathematical subjects. The *x*- and *y*-coordinates (the values that specify a location on a graph) are called *Cartesian coordinates* in honor of Descartes.

Plotting X versus Y

Plotting points in a coordinate system involves counting distances to the right or left and up or down from the origin. The axes serve as a starting place. The points are represented or named by the ordered pair of numbers, (x,y), called the *x*-coordinate and the *y*-coordinate. The designation *ordered pair* means that the order *does* matter. The *x*-coordinate always comes first, and the *y*-coordinate comes last so that this whole graphing system is universal.

The *x*-coordinate is the distance to the left or right of the origin that the point lies. If the *x*-coordinate is positive, you move to the right of the origin. If it's negative, you move to the left. The second number, the *y*-coordinate, is the distance up or down from the origin. Positive numbers mean the point is up, and negatives mean you move south of the *x*-axis.

The point $(2,4)$ is two units to the right and four units up from the origin; $(-3,2)$ is three units to the left and two units up; $(-4,-3)$ is four units to the left and three units down; and $(5,-1)$ is five units to the right and one unit down. Points can lie on one of the axes, too. Those points always have a 0 for the *x*- or *y*-coordinate. The point $(0,3)$ lies on the *y*-axis, and $(1,0)$ lies on the *x*-axis. See how to graph all these points in Figure 2-2.

Cutting the graph into four parts

The intersecting *x*- and *y*-axes divide the whole picture, or *coordinate plane*, into four separate regions called *quadrants*. The quadrants are numbered, starting in the upper right quadrant and going counterclockwise, as shown in Figure 2-3. Traditionally, you number them with Roman numerals.

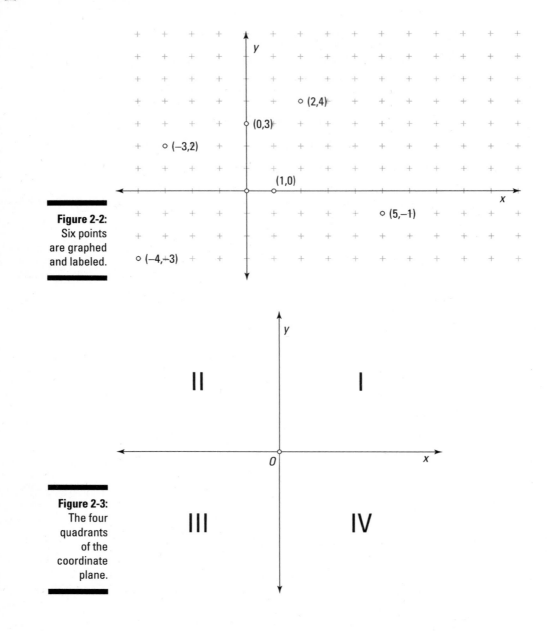

Figure 2-2:
Six points are graphed and labeled.

Figure 2-3:
The four quadrants of the coordinate plane.

These quadrant number designations are useful when referring to certain types of angles and groupings of points. The points in Quadrant I all have both *x*- and *y*-coordinates that are positive. In Quadrant II, the *x*-coordinate is negative, and the *y*-coordinate is positive. The points in Quadrant III have both *x*- and *y*-coordinates that are negative. In Quadrant IV, the *x*-coordinate is positive, and the *y*-coordinate is negative.

From Here to There: Calculating Distances

The lengths of segments and distances between points play a major part in establishing the trig functions, relationships, and identities (which I cover in Chapters 3 and 10). You can compute these lengths and distances fairly easily, because the coordinate system is just so darned convenient.

Counting on vertical and horizontal distances

When the distance that you're measuring is either vertical or horizontal, then the computation is a simple subtraction problem. One coordinate in each ordered pair is the same. Just find the difference between the other two coordinates.

For instance, to find the distance between the points (5,2) and (5,6), subtract 2 from 6 to get the distance of 4 units between them. This distance is vertical, because the two points have the same *x*-coordinate, and the second point is directly above the first. To find the distance between the two points (5,6) and (5,–3), subtract –3 from 6 to get a distance of 9 units. Always subtract the smaller number from the larger number, so that the distance you get is a positive number. (Negative distances don't make sense; after all, you can't travel –5 miles to Aunt Myrtle's house!)

Another way to deal with the different signs of the answers that occur is to use absolute value — then it doesn't matter in what order you subtract the numbers. Take, for example, the preceding example, where I subtracted –3 from 6. The –3 is smaller, so subtracting in that order gave a positive answer. The alternative is to subtract in the opposite order and take the absolute value of the result. If you do –3 – 6, you get –9. The absolute value of –9, written |–9|, equals 9.

Horizontal distances work the same way. In Figure 2-4, you can see the horizontal distance between two points. To find the distance between the points (–8,2) and (5,2), just calculate the difference between –8 and 2, because the y-coordinates are the same. The smaller number is –8, so subtracting $5 - (-8)$, the answer is 13 units. Subtracting in the other order and using absolute value, you get $-8 - 5 = -13$, and $|-13| = 13$. You can also see the vertical distance between two points, (5,6) and (5,2), in Figure 2-4. This problem uses simple arithmetic. The difference between 6 and 2 is 4.

Figure 2-4:
Vertical and horizontal distances between points.

Finding the distance between two different vertical or horizontal points, (x_1, y_1) and (x_2, y_2), is easy:

- ✔ Vertical distance (the x-coordinates are the same) is $|y_1 - y_2|$.
- ✔ Horizontal distance (the y-coordinates are the same) is $|x_1 - x_2|$.

Another slant: Diagonal distances

Sometimes the distances or lengths you want to determine are on a slant — they go diagonally from one point to another. The formula for determining these distances is based on the Pythagorean theorem.

The Pythagorean theorem

Way back when, Pythagoras discovered a relationship between the sides of any *right triangle,* where one angle is 90 degrees, as Figure 2-5 shows.

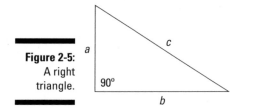

Figure 2-5:
A right
triangle.

Pythagoras found that if a and b are the lengths of the shorter sides of the right triangle, and if c is the length of the *hypotenuse* (the side opposite the right angle), then $a^2 + b^2 = c^2$. You can use this formula to find the diagonal distances between two points on a graph, because the horizontal and vertical distances, which are the sides of the triangle, are easy to find in a coordinate system.

Determining diagonal distances

Using the Pythagorean theorem, solve for c, the length of the hypotenuse of a right triangle, and you get

$$c = \sqrt{a^2 + b^2}$$

If the length a is the horizontal distance, then you calculate that distance by subtracting the x-coordinates; if length b is the vertical distance, you get it by subtracting the y-coordinates.

To get the general distance formula, simply substitute the difference between the x- and y-values for a and b in the Pythagorean theorem, and use the variable d (meaning distance) in place of c.

The distance, d, between two points (x_1, y_1) and (x_2, y_2) is

$$d = \sqrt{(x_1 - x_2)^2 + (y_1 - y_2)^2}$$

For example, follow these steps to find the distance between the points $(3, -4)$ and $(-2, 5)$:

1. **Replace x_1 and x_2 with 3 and –2. Replace the y_1 and y_2 with –4 and 5.**

   $$d = \sqrt{(3 - (-2))^2 + (-4 - 5)^2}$$

2. **Subtract the coordinates.**

   $$d = \sqrt{5^2 + 9^2}$$

3. **Add the results and find the square root, if possible.**

   $$d = \sqrt{25 + 81} = \sqrt{106}$$

In the preceding example, the number under the radical isn't a perfect square. You can either leave the answer with the square-root symbol or give a decimal approximation (see the following section). To three decimal places, the distance in this example is 10.296 units.

When you're calculating the distance between two points, it doesn't matter in what order you subtract the points, as long as you subtract x from x and y from y. Squaring the differences will always result in a positive answer, anyway.

Using exact values or estimating distances

Calculating the distance between two points often leaves you with the square root of a number that isn't a perfect square; this type of answer is called an *irrational number*. Writing the number with the square-root symbol, for example, $\sqrt{47}$, is considered to be writing the *exact value* of the distance. Using a calculator to find a decimal approximation doesn't give an exact answer, because the decimal values go on forever and ever and never repeat in a pattern. Because the decimals are always estimates, mathematicians often insist on leaving the answers as exact values, complete with the square-root symbol, rather than decimals.

Although exact values are more precise, using decimal estimates of radical values is more helpful in practical situations. If you're solving for the height of a building and get $\sqrt{183}$, you get a better understanding of the height by finding a decimal estimate. A scientific calculator tells you that $\sqrt{183}$ is approximately 13.52774926. . . . Different calculators may give you fewer or more decimal values than I show you here. Usually, just two or three decimal places will do. Rounding this to two places, you get 13.53. Rounding it to three places, you get 13.528. (For more on rounding, please refer to Chapter 1.)

Getting to the Center of It All

The middle of a circle is its *center*. The middle of a triangle is its *centroid*. And the middle of a line segment is its *midpoint*. When you graph a circle, triangle, or line segment by using coordinate axes, then you can name these middle points with a pair of x- and y-coordinates. All you need to find these middles are the coordinates of some crucial other points on the respective figures.

Finding the midpoint of a line segment

To find the midpoint of a line segment, you just calculate the averages of the coordinates — easy as pie.

TRIG RULES
$$\frac{\begin{array}{r} 1 \\ +\,1 \end{array}}{2}$$

The midpoint, M, of a segment with endpoints (x_1, y_1) and (x_2, y_2) is

$$M = \left(\frac{x_1 + x_2}{2}, \frac{y_1 + y_2}{2} \right)$$

If you want to know the midpoint of the segment with endpoints $(-4, -1)$ and $(2, 5)$, then plug the numbers into the midpoint formula, and you get a midpoint of $(-1, 2)$:

$$M = \left(\frac{-4 + 2}{2}, \frac{-1 + 5}{2} \right) = \left(\frac{-2}{2}, \frac{4}{2} \right) = (-1, 2)$$

See how this segment looks in graph form in Figure 2-6.

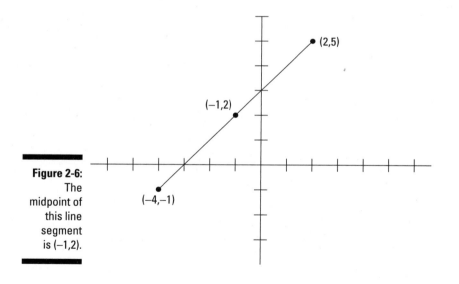

Figure 2-6: The midpoint of this line segment is $(-1, 2)$.

Locating the center of a circle

If the endpoints of one diameter of a circle are (x_1, y_1) and (x_2, y_2), then the center of the circle has the coordinates $\left(\frac{x_1 + x_2}{2}, \frac{y_1 + y_2}{2} \right)$. You probably noticed that the center of a circle is the same as the diameter's midpoint. The center of the circle separates the diameter into two equal segments called *radii* (plural for radius).

Figure 2-7 shows a circle with a diameter whose endpoints are (7,4) and (–1,–2). The center of the circle is at (3,1). I got the coordinates for the center by using the formula for the midpoint of a segment (see the preceding section):

$$M = \left(\frac{7 + (-1)}{2}, \frac{4 + (-2)}{2} \right) = \left(\frac{6}{2}, \frac{2}{2} \right) = (3, 1)$$

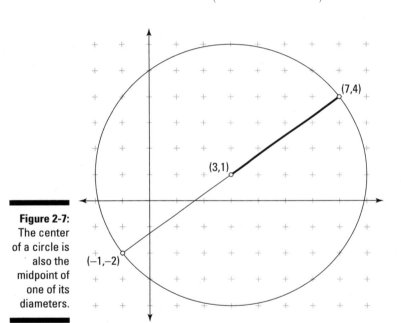

Figure 2-7:
The center of a circle is also the midpoint of one of its diameters.

You find the length of the diameter by using the distance formula (see the section "Another slant: Diagonal distances," earlier in this chapter):

$$d = \sqrt{\left(7 - (-1)\right)^2 + \left(4 - (-2)\right)^2} = \sqrt{8^2 + 6^2} = \sqrt{64 + 36} = \sqrt{100} = 10$$

For the circle shown in Figure 2-7, the diameter is 10 units long.

Next, I show you how to find the length of one of the radii. Either will do — they're the same length. In this example, I figure the radius length from the center of the circle (3,1) to the endpoint of the diameter (7,4):

$$d = \sqrt{(7-3)^2 + (4-1)^2} = \sqrt{4^2 + 3^2} = \sqrt{16+9} = \sqrt{25} = 5$$

The radius is 5 units long. But, of course, you expected this answer, because by definition, the radius is half the length of the diameter.

Partitioning line segments further

If you can find the midpoint of a segment, then you can divide it into two equal parts. Finding the middle of each of the two equal parts allows you to find the points needed to divide the entire segment into four equal parts. Finding the middle of each of these segments gives you eight equal parts, and so on.

For example, to divide the segment with endpoints (–15,10) and (9,2) into eight equal parts, find the various midpoints like so:

- ✔ The midpoint of the main segment from (–15,10) to (9,2) is (–3,6).

- ✔ The midpoint of half of the main segment, from (–15,10) to (–3,6), is (–9,8), and the midpoint of the other half of the main segment, from (–3,6) to (9,2), is (3,4).

- ✔ The midpoints of the four segments determined above are (–12,9), (–6,7), (0,5), and (6,3).

Figure 2-8 shows the coordinates of the points that divide this line segment into eight equal parts.

Using the midpoint method is fine, as long as you just want to divide a segment into an even number of equal segments. But your job isn't always so easy. For instance, you may need to divide a segment into three equal parts, five equal parts, or some other odd number of equal parts.

To find a point that isn't equidistant from the endpoints of a segment, just use this formula:

$$(x,y) = \left(x_1 + k(x_2 - x_1), y_1 + k(y_2 - y_1)\right)$$

In this formula, (x_1, y_1) is the endpoint where you're starting, (x_2, y_2) is the other endpoint, and k is the fractional part of the segment you want.

So, to find the coordinates that divide the segment with endpoints (–4,1) and (8,7) into three equal parts, first find the point that's ⅓ of the distance from (–4,1) to the other endpoint, and then find the point that's ⅔ of the distance from (–4,1) to the other endpoint. The following steps show you how.

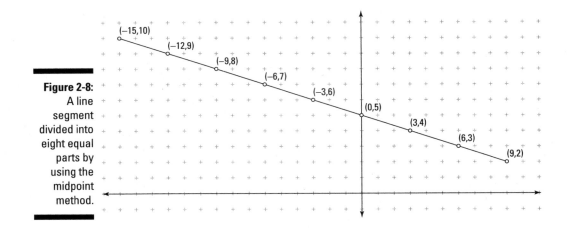

Figure 2-8:
A line
segment
divided into
eight equal
parts by
using the
midpoint
method.

To find the point that's ⅓ of the distance from (–4,1) to the other endpoint, (8,7):

1. **Replace x_1 with –4, x_2 with 8, y_1 with 1, y_2 with 7, and k with $\frac{1}{3}$.**

$$(x,y) = \left(-4 + \frac{1}{3}(8 - (-4)), 1 + \frac{1}{3}(7 - 1)\right)$$

2. **Subtract the values in the inner parentheses.**

$$(x,y) = \left(-4 + \frac{1}{3}(12), 1 + \frac{1}{3}(6)\right)$$

3. **Do the multiplication and then add the results to get the coordinates.**

$$(x,y) = (-4 + 4, 1 + 2) = (0,3)$$

To find the point that's ⅔ of the distance from (–4,1) to the other endpoint, (8,7):

1. **Replace x_1 with –4, x_2 with 8, y_1 with 1, y_2 with 7, and k with $\frac{2}{3}$.**

$$(x,y) = \left(-4 + \frac{2}{3}(8 - (-4)), 1 + \frac{2}{3}(7 - 1)\right)$$

2. **Subtract the values in the inner parentheses.**

$$(x,y) = \left(-4 + \frac{2}{3}(12), 1 + \frac{2}{3}(6)\right)$$

3. **Do the multiplication and then add the results to get the coordinates.**

$$(x,y) = (-4 + 8, 1 + 4) = (4,5)$$

Figure 2-9 shows the graph of this line segment and the points that divide it into three equal parts.

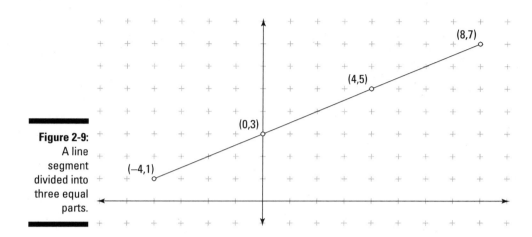

Figure 2-9:
A line
segment
divided into
three equal
parts.

Pinpointing the center of a triangle

If you draw lines from each corner (or *vertex*) of a triangle to the midpoint of the opposite sides, then those three lines meet at the center, or *centroid,* of the triangle. This point is the triangle's center of gravity, where the triangle balances evenly. The coordinates of the centroid are also ⅔ of the way from each vertex along that segment. Figure 2-10 shows how the three lines drawn in the triangle all meet at the center.

Figure 2-10:
The
lines that
intersect at
the centroid
of a triangle.

To find the centroid of a triangle, use the formula from the preceding section that locates a point ⅔ of the distance from the vertex to the midpoint of the opposite side.

Circumscribing a triangle

Every triangle can be *circumscribed* by a circle, meaning that one circle — and only one — goes through all three *vertices* (corners) of any triangle. In laymen's terms, any triangle can fit into some circle with its corners touching the circle. To circumscribe a triangle, all you need to do is find the center of the circle. You can then find the radius of the circle, because the distance from the center of the circle to one of the triangle's vertices is the radius. This exercise is a nice one to try your hand at with a compass and straightedge or with some of the new geometry software.

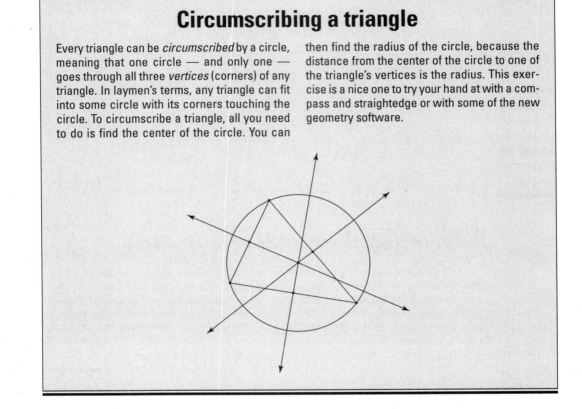

For example, to find the centroid of a triangle with vertices at (0,0), (12,0) and (3,9), first find the midpoint of one of the sides. The most convenient side is the bottom, because it lies along the *x*-axis. The coordinates of that midpoint are (6,0). Then find the point that sits ⅔ of the way from the opposite vertex, (3,9):

1. **Replace x_1, x_2, y_1, and y_2 with their respective values. Replace k with $\frac{2}{3}$.**

$$(x,y) = \left(3 + \frac{2}{3}\big(6 - (3)\big), 9 + \frac{2}{3}(0 - 9)\right)$$

2. **Simplify the computation to get the point.**

$$= \left(3 + \frac{2}{3}(3), 9 + \frac{2}{3}(-9)\right) = (3 + 2, 9 - 6) = (5,3)$$

In this example, the centroid is the point (5,3), as shown in Figure 2-11.

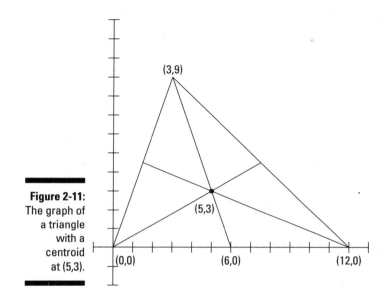

Racing Down the Slope

In mathematics, a *slope* is a particular number or value that tells you something about the nature of a line or line segment. Just by looking at the number corresponding to the slope of a line, you can tell if the line goes up or down as you read from left to right. You can also tell if the slope of the line is steep or rather flat (like the slopes in Colorado versus those in Illinois).

Slaloming slope formula

One way to find the slope of a line or segment is to choose any two points, (x_1,y_1) and (x_2,y_2), on the figure and use the formula that gives you the slope, represented by the letter *m*:

$$m = \frac{y_2 - y_1}{x_2 - x_1}$$

For example, the slope of the line through the points (–2,2) and (1,8) is

$$m = \frac{2 - 8}{-2 - 1} = \frac{-6}{-3} = 2$$

This line moves upward from left to right, which is why the slope is a positive number. Any slope greater than 1 is also considered to be *steep*.

On the other hand, the slope of the line through the points (–5,2) and (5,–1) is

$$m = \frac{2 - (-1)}{-5 - 5} = \frac{3}{-10} = -\frac{3}{10}$$

This segment moves downward from left to right, and the unsigned value (absolute value) of the slope is a number between 0 and 1 — so the slope is negative, and it isn't considered steep.

Figure 2-12 shows both lines, one with a slope of 2 and the other with a slope of $-\frac{3}{10}$.

Recognizing parallel and perpendicular lines

Two lines are *parallel* if they have the same slope. Two lines are *perpendicular* if their slopes are negative reciprocals of one another. By being *negative reciprocals,* when you multiply the slopes, you get –1.

Consider the following slopes of some lines or line segments:

$$m_1 = \frac{1}{2}, \ m_2 = -2, \ m_3 = \frac{6}{3}, \ m_4 = \frac{5}{10}, \ m_5 = -\frac{14}{7}$$

The slopes of the lines that are parallel to one another are

$$m_1 = \frac{1}{2} \text{ and } m_4 = \frac{5}{10}$$
$$m_2 = -2 \text{ and } m_5 = -\frac{14}{7}$$

The slopes of lines that are perpendicular to one another are

$$m_1 = \frac{1}{2} \text{ and } m_2 = -2$$
$$m_4 = \frac{5}{10} \text{ and } m_5 = -\frac{14}{7}$$

As a matter of fact, because the lines with slopes of $\frac{1}{2}$ and $\frac{5}{10}$ are equal to one another, they're both perpendicular to the lines with slopes of –2 and $-\frac{14}{7}$, which are also equal in slope. It's one big, happy family.

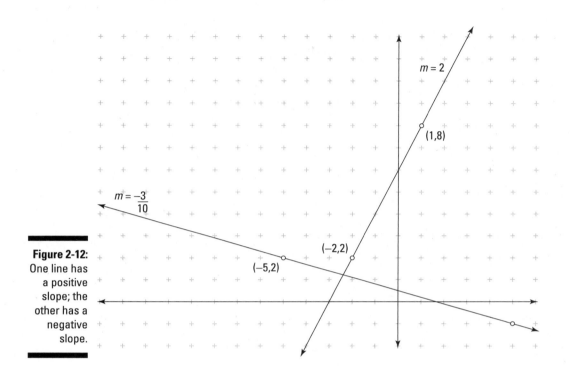

Figure 2-12:
One line has
a positive
slope; the
other has a
negative
slope.

Defining Circles with Numbers

The circle that you use the most in trigonometry has its center at the origin and has a radius of 1 unit. The radius of 1 in a circle makes computations so much easier when that 1 ends up in the denominator of a fraction. Fractions and circles sort of intermingle in trigonometry — in good ways, of course. But you also have many other useful circles to consider. The other circles will have different radii and different centers, but each has its place when needed. When possible, though, the unit circle is the circle of choice.

Centering circles at the origin

The two characteristics that define a circle are its center and its radius. The center tells where on a graph the circle is located; the radius tells how big the circle is. The location is in terms of coordinates in the coordinate plane, and those numbers end up in the equation of the circle. The x and y variables represent the coordinates of all the points that lie on the actual circle. The standard form for the equation of a circle at the origin is $x^2 + y^2 = r^2$, where r represents the radius of the circle. So the equations for circles with radii of 2, 3, 4, and 5 are $x^2 + y^2 = 4$, $x^2 + y^2 = 9$, $x^2 + y^2 = 16$, and $x^2 + y^2 = 25$, respectively.

Likewise, a circle with a radius of 1 unit has the equation $x^2 + y^2 = 1$. Not so amazingly, this circle is called the *unit circle*. This circle is used extensively in mathematics: The radius of 1 lends itself to the formula for changing from degrees to radians, is nice and neat when finding arc length, and makes the unit circle the easiest to use when proving properties or theorems in math.

Wandering centers

Circles don't have to have their centers at the origin. The standard form for a circle with a radius of r and its center at (h,k) is $(x-h)^2 + (y-k)^2 = r^2$, where x and y represent the coordinates of all the points on that circle. So the equation for a circle with its center at (3,–2) and with a radius of 9 is $(x-3)^2 + (y+2)^2 = 81$.

Notice that, if you let the center of a circle be (0,0) in this formula, you get $(x-0)^2 + (y-0)^2 = r^2$ or $x^2 + y^2 = r^2$, which goes back to a circle with its center at the origin. The form works for all circles!

Chapter 3

Functioning Well

In This Chapter

▶ Understanding why functions are your friends

▶ Applying the inverse to a function

▶ Moving a function around on a graph

*Y*ou can't get very far in any mathematical discussion without encountering rules, patterns, operations, or relationships among the concepts you're discussing. One common theme in math is the relationship between certain values (often called the *input* and the *output*), which are the values you start with and the values you end up with, respectively. *Functions* are very special types of relationships, and they play a big part in trigonometry. So what is it that distinguishes a relation from a function, and why should you care? The distinction is important in all of mathematics, not just in trigonometry.

Relations versus Functions

A *relation* in mathematics is a rule that creates a certain output for any given input. The *input* is the number you enter in place of a variable, and the *output* is the result(s) you get when you perform the operations for that relation. Each relation has a rule, or expression, that usually involves mathematical operations such as addition, subtraction, square roots, and so on. For instance, you could come across a relation where you input 25 in place of a variable and get two output values, such as 24 and 27. The rule for that relation could be that you input a number and get the two numbers closest to it that are multiples of 3. This relation has more than one output value, which isn't necessarily a good thing; in math, more isn't always better.

A *function,* however, is a special kind of relation. Read on to find out more.

Function junction, what's your function?

A function in mathematics is a rule performing operations and processes on input values that results in a single, unique output value — only one output for every input. For example, take the function where the input is the number 25 and the output is 5. Now you have several ways of getting the result of 5 after inputting 25 in place of a variable: You can take the square root of the number or subtract 20 from the number, for starters. But the main emphasis here is that the function has just *one* output value.

Consider a function that uses a *radical* (a root). Input 25, and here's what that function looks like: $\sqrt{25} = 5$. The output is the single value 5. Another function is one that squares the input, multiplies that result by 2, and then subtracts 3. If you input an 8, then you get $2x^2 - 3 = 2 \cdot 8^2 - 3 = 2 \cdot 64 - 3 = 125$. The preceding examples of functions use the basic algebraic operations. But I'm here to tell you that a whole class of functions called *trigonometric functions* is out there, too. That's why you're reading this book! One of the trig functions is called *sine,* abbreviated *sin.* If you compute the sine of 30 degrees, then you get $\frac{1}{2}$. Because sine is a function, $\frac{1}{2}$ is the only output value. It may seem trivial or unnecessary for me to keep harping on that right now, but having only one output value for each input in trigonometry is very important — otherwise, you'd have chaos!

Using function notation

Defining a function or explaining how it works can involve a lot of words and can get rather lengthy and awkward. Imagine having to write, "Square the input, multiply that result by 2, and then subtract 3." Mathematicians are an efficient lot, and they prefer a more precise, quick way of writing their instructions. Function notation is just that.

First, functions are generally named with letters — the most frequently used is the oh-so-obvious *f.* (I said that mathematicians are efficient, but they're not necessarily original or creative.) If I want the function *f* to be the rule for squaring a number, multiplying that result by 2, and then subtracting 3, I write the function as $f(x) = 2x^2 - 3$. You read the function like this: "*F* of *x* is equal to two times *x* squared minus 3." The *x* is a variable — in this case, the input variable. Whatever you put in the parentheses after the *f* replaces any *x* in the rule. In the first equation that follows, an 8 replaces the *x*. In the second equation, a −4 replaces the *x*. Each time, the function produces only one answer:

$$f(8) = 2 \cdot 8^2 - 3 = 128 - 3 = 125$$
$$f(-4) = 2(-4)^2 - 3 = 32 - 3 = 29$$

Don't feel bound to the *f*, though. You can use other letters to name the functions and the input variables. Sometimes you use letters that represent what's going on or what you're using the formula for, such as finding area, interest, or cost:

$$A(r) = \pi r^2$$
$$I(t) = 1000e^{0.04t}$$
$$C(x) = -0.04x^2 + 8x + 100$$

And, of course, you have the trig functions. Some trig functions involving sine, cosine, and secant are

$$p(x) = \sin x + \cos x$$
$$c(\theta) = \frac{1}{\sec \theta}$$

Determining domain and range

A function consists of a rule that you apply to the input values. The result is a single output value. You can usually use a huge number of input values, and they're all a part of the *domain* of the function. The output values make up the *range* of the function.

Are you master of your domain?

The domain of a function consists of all the values that you can use as input into the function rule. The domain is another of that function's characteristics, because different functions have different numbers that you can input and have the outputs make any sense.

For example, $f(x) = \sqrt{x}$ is a function whose domain can't contain any negative numbers, because the square root of a negative number isn't a real number. The function $g(x) = \frac{4}{x+3}$ has a domain that can't include the number –3. Any other real number is okay, but not –3, because putting a –3 in for *x* makes the denominator equal to 0, and you can't divide by 0. (A fraction with a 0 in the denominator doesn't exist.) With trig functions, the input values are always angle measures — either in degrees or radians. Some of the trig functions have restrictions, too. For example, the tangent function has a domain that can't include 90 degrees or 270 degrees, among the many restricted values. I discuss these domains in detail in Chapter 7.

Home, home on the range

The range of a function consists of all its output values — the numbers you get when you input numbers from the domain into the function and perform the function operations on them. Sometimes, a range can be all possible real numbers — it has no limit. That situation happens in a function such as $h(x) = 3x + 2$. In this equation, both the domain and the range are unlimited. You can put in any real number, and you can get an output of any real number that you can possibly think of. Ranges can end up being restricted, though. For example, the function $k(x) = x^2 + 6$ will always have results that are either the number 6 or some positive number greater than 6. You can never get a negative number or a number less than 6 as an output. The ranges of some trig functions are restricted, too. For example, the output of the sine function never exceeds 1 or goes lower than –1. I cover this subject in detail in Chapter 7.

In-Verse Functions: Rhyme or Reason?

Functions are special types of relationships between mathematical values, because they yield only one unique output value for every input value. For a more-detailed definition of a function, see the "Function junction, what's your function?" section, earlier in this chapter. Sometimes, you have to work backwards with functions, because you know the output value and you want to figure out which input value gave you that output. That's where *inverse functions* come in.

Which functions have inverses?

The best way to describe an inverse function is to give an example: one formula or function tells you what the temperature outside is in degrees Celsius when you input a temperature in degrees Fahrenheit. The function is $C(f) = \frac{5}{9}(f - 32)$, where C is the answer in Celsius degrees and f is the temperature in Fahrenheit degrees. If you input 77 degrees into the function, you get $C(77) = \frac{5}{9}(77 - 32) = \frac{5}{9}(45) = 25$, or 25 degrees Celsius.

Now, how about going the other way? What if you want to know what temperature in Celsius will *give you* 77 degrees Fahrenheit? You take the inverse of the function, which I show you how to do in the next section. But stick with me for now. Keep in mind that only one temperature gives you that answer of 77 degrees. (It wouldn't make sense if both 25 degrees and 45 degrees Celsius, for example, gave the same answer of 77 degrees Fahrenheit.)

So you see the main distinction between functions that have inverses and functions that don't: A function can have an inverse function only if the function is *one-to-one*. In other words, the function has to be designed in such a way that every input has exactly one output *and* every output comes from only one input — the output doesn't occur with more than one input value.

An example of a function that has an inverse function is $f(x) = 4x + 5$. Its inverse is

$$f^{-1}(x) = \frac{x-5}{4}$$

Notice that the function notation for the name of the inverse function is the same letter but with a –1 exponent. This exponent doesn't mean that you want a reciprocal; rather, the –1 exponent in a function name is special math notation meaning an inverse function. You see this notation a lot in trigonometry. The inverse functions all have a name using the –1 exponent after the corresponding function name. They can have an alternative name, too. For more on this naming mumbo-jumbo, go to Chapter 13.

Check out how this inverse function works by using the last function I showed you. If you input a 6 into the function, f, you get $f(6) = 4 \cdot 6 + 5 = 29$. Take that output, 29, and put it into the inverse function to see where that particular output came from: $f^{-1}(29) = \frac{29-5}{4} = 6$. This function and its inverse are one-to-one. No other input into f will give you 29, and no other input into f^{-1} will give you 6.

Not all functions have inverses, though. An example of a function that is *not* one-to-one is $g(x) = x^2 - 4$. If you input 7, you get $g(7) = 7^2 - 4 = 45$. But you also get 45 if you input –7: $g(-7) = (-7)^2 - 4 = 49 - 4 = 45$. From the output value, you can't possibly tell which of the two numbers was the input value. This function isn't one-to-one, so it doesn't have an inverse.

Sometimes, you can spot a function that has an inverse, and sometimes that quality isn't so apparent. Here are some fairly obvious clues that you can pick up on just by looking at the rule for the function. A function does *not* have an inverse if

✔ The function rule has an even exponent (not including 0).

✔ The function rule has an absolute value symbol.

✔ The graph of the function is a horizontal line.

✔ You draw a horizontal line through the graph of the function and that line intersects the graph more than once.

The trig functions all have inverses, but only under special conditions — you have to restrict the domain values. I discuss what it means to restrict the domain values in Chapter 13.

Finding an inverse function

Not all functions have inverses, and not all inverses are easy to determine. Here's a nice method for finding inverses of basic algebraic functions.

Using algebra

The most efficient method for finding an inverse function for a given one-to-one function involves the following steps:

1. **Replace the function notation name with y.**

2. **Reverse all the x's and y's (let every x be y and every y be x).**

3. **Solve the equation for y.**

4. **Replace y with the function notation for an inverse function.**

For example, to find the inverse function for $f(x) = \sqrt[3]{x-2} + 8$:

1. **Replace the function notation with y.**

   $y = \sqrt[3]{x-2} + 8$

2. **Reverse the x's and y's.**

   $x = \sqrt[3]{y-2} + 8$

3. **Solve for y.**

   $$x - 8 = \sqrt[3]{y-2}$$
   $$(x-8)^3 = y - 2$$
   $$(x-8)^3 + 2 = y$$

4. **Replace y with the inverse function notation.**

   $f^{-1}(x) = (x-8)^3 + 2$

Look at how these two functions work. Input 3 into the original function and then get the number 3 back again by putting the output, 9, into the inverse function.

1. **Replace the x's with 3 in the function.**

   $f(3) = \sqrt[3]{3-2} + 8 = \sqrt[3]{1} + 8 = 9$

2. **Replace the x's with 9 in the inverse function.**

   $f^{-1}(9) = (9-8)^3 + 2 = 1^3 + 2 = 3$

Using new definitions of functions for inverses

Sometimes you just don't have a nice or convenient algebraic process that will give you an inverse function. Many functions need a special, new rule for their inverse. Some examples of these functions are

Function	**Inverse**
$f(x) = e^x$	$f^{-1}(x) = \ln x$
$g(x) = \log_a x$	$g^{-1}(x) = a^x$
$h(x) = \sin x$	$h^{-1}(x) = \arcsin x$ or $\sin^{-1} x$
$k(x) = \tan x$	$k^{-1}(x) = \tan^{-1} x$ or $\arctan x$

If you have a scientific or graphing calculator, you can try out some of these functions and their inverses. Use the function $f(x) = e^x$ and its inverse, $f^{-1}(x) = \ln x$, for the following demonstration.

1. **In the calculator, use the e^x button (often a 2nd function of the calculators) to enter e^3.**

   The input value here is 3. The answer, or output, comes out to be about 20.08553692. This value isn't exact, but it's good for eight decimal places.

2. **Now take that answer and use the *ln* button to find *ln* 20.08553692.**

   Input 20.08553692 into the *ln* function. The answer, or output, that you get this time is 3.

Transforming Functions

Functions and all their properties, characteristics, and peculiarities are of interest to mathematicians and others who use them as models for practical applications. Using functions to find values or answers to practical problems is helpful only if tweaking or slightly changing the functions is reasonably simple. Predictable and controlled changes of functions meet this requirement of ease and simplicity. Chapter 19 deals with how the transformations affect trig functions. This section gives you a more-general explanation of how to tweak your functions.

Translating a function

A translation is a slide, which means that the function has the same shape graphically, but the graph of the function slides up or down or slides left or right on the coordinate plane.

Sliding up or down

Figure 3-1 shows the parabola $y = x^2$ with a translation 5 units up and a translation 7 units down. A *parabola* is the graph of a second-degree polynomial, which means that the polynomial has a power of 2 for one exponent. The graph makes a nice, *U*-shaped curve.

Think about a function that you use to determine how much money a person earns for working a certain number of hours. The amount can slide up or down if you add a bonus or subtract a penalty from the amount. Here's what the situation may look like in function notation:

- Translating up C units: $f(x) + C$
- Translating down C units: $f(x) - C$

A person who makes $8 an hour but gets a $50 bonus has a pay function that looks like $P(h) = 8h + 50$. If that same person were penalized $6 for being late, the pay function would look like $P(h) = 8h - 6$.

Sliding left or right

Figure 3-2 shows the parabola $y = x^2$ with a translation 5 units right and a translation 7 units left.

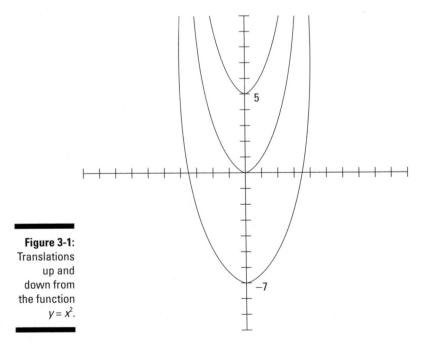

Figure 3-1:
Translations up and down from the function $y = x^2$.

5

−7

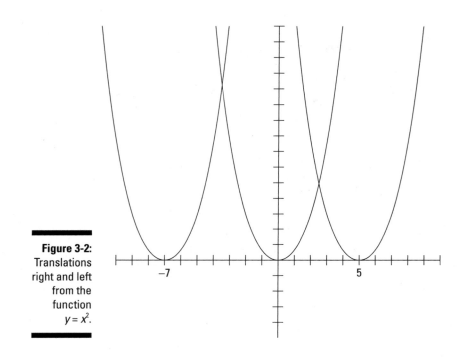

If you use a function to determine how much commission a person earns for selling a certain number of computers, the commission can be affected when you add or subtract the number of units the person needs to sell. Here's what the situation looks like in function notation:

✔ Translating left C units: $f(x + C)$

✔ Translating right C units: $f(x - C)$

A person who makes $50 commission for every computer sold but gets upfront credit for two computers as an incentive has a commission function that looks like $P(x) = 50(x + 2)$. On the other hand, a person who has the same commission schedule but had two computers returned and starts with a deficit has a commission function that looks like $P(x) = 50(x - 2)$.

Reflecting like a mirror

Two types of transformations act like reflections or flips. One transformation changes all positive outputs to negative and all negative outputs to positive. The other reverses the inputs — positive to negative and negative to positive.

✔ Reflecting up and down (outputs changed): $-f(x)$

✔ Reflecting left and right (inputs changed): $f(-x)$

Figure 3-3 shows reflections of the function $y = \sqrt{x}$. Reflecting downward puts all the points below the x-axis. Reflecting left makes all the input values move to the left of the y-axis. Even though it appears that the negatives should go under the radical, in fact, the negative in front of the x means that you take the opposite of all the negative x's — which makes them positive.

A cash register can change inputs to the opposite (negative) numbers by taking coupon values that the cashier enters or scans in and changing them to negative values before doing the final computations. The graph of this process acts as a reflection downward from positive to negative.

Left and right reflections are a bit harder to describe in terms of a practical application. Try this one on for size: If a function tells you how many items a machine can produce in a certain number of hours, then inputting negative numbers helps you determine how far you have to back up — how many hours before a certain date and time — to produce that number of items by that date and time.

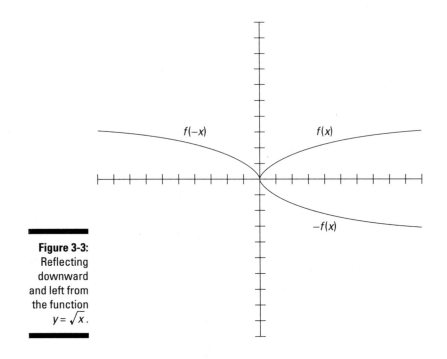

Figure 3-3:
Reflecting
downward
and left from
the function
$y = \sqrt{x}$.

Chapter 4

Getting Your Degree

*T*he main idea that distinguishes trigonometry from other mathematical topics is its attention to and dependence on angle measures. The trig functions (sine, cosine, tangent, cotangent, secant, and cosecant) are ratios based on the measures of an angle. What good are degrees (no, not the kind that tell you how hot or cold it is) in the real world? Navigators, carpenters, and astronomers can't do without them. How do you measure the degrees? You have many ways, dear reader, and I show you all you need to know in this chapter.

Angles, Angles Everywhere: Measuring in Degrees

What's a degree? When you graduate from college, you get your degree. The temperature outside went up a degree. When questioned, you get the third degree. All these scenarios use the word *degree,* but in trigonometry, a degree is a tiny slice of a circle. Imagine a pizza cut into 360 equal pieces (what a mess). Each little slice represents one degree. Look at Figure 4-1 to see what a degree looks like.

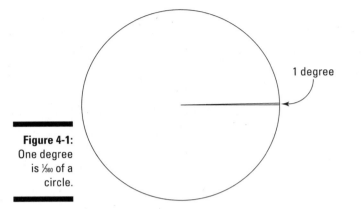

Figure 4-1:
One degree
is ⅟₃₆₀ of a
circle.

Slicing a coordinate plane

The first quadrant is the upper right-hand corner of the coordinate plane. (See Chapter 2 for the lowdown on quadrants and coordinate planes.) That first quadrant is ¼ of the entire plane. So if a full circle with its center at the origin has a total of 360 degrees, then ¼ of it has 90 degrees, which is the measure of the angle that the first quadrant forms. Actually, each quadrant measures exactly 90 degrees. You can divide each of these 90-degree measures evenly by many numbers, and you use those equal divisions frequently in trig, because they're nice, neat divisions. The most frequently used angle measures include $\frac{90}{2} = 45$ degrees, $\frac{90}{3} = 30$ degrees, and $\frac{90}{6} = 15$ degrees. And then, twice the 30-degree angle is 60 degrees (another common angle in trig).

This elite group of angle measures is 0, 15, 30, 45, 60, and 90 degrees. These angles and their multiples occupy much of the discussion in trigonometry because of their convenience in computations. Figure 4-2 shows sketches of some of the angles.

Figure 4-2:
Some of
the most
commonly
used angles:
90, 60, 45,
30, and 15
degrees.

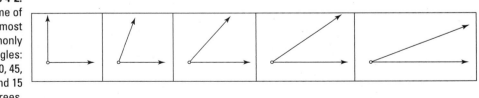

Looking elsewhere for degree measures

Your first introduction to the idea of measuring angles in degrees probably didn't come from a course in geometry or trigonometry. Most of us are exposed to this idea through television or movies. A popular situation in such shows involves a plane flying through a storm or at night or with no one but a stewardess at the controls. A radio transmission from the control tower comes crackling through all the static, with an announcer saying, "Turn to a heading of 40 degrees." And because the pilot or stewardess remembers her trigonometry, she saves the day. Hurray for degrees!

Another type of situation that you find on television is on _The New Yankee Workshop_ starring Norm Abram, where Norm, in all his woodworking grandeur, cuts a board at a 30-degree angle so it fits perfectly in a carefully crafted wooden bookcase.

Navigating with degrees

In navigation and surveying, the _bearing_ or _heading_ is the direction that a plane, boat, or line takes. In math speak, this bearing is the angle measured in degrees that a _ray_ (a line with one endpoint that extends infinitely in the other direction) makes with a second ray that points north. The angle is measured in a clockwise direction. (Note, however, that in the _standard position_ in geometry and trigonometry, you measure angles in a _counterclockwise_ direction.) Figure 4-3 shows some bearings. Notice that the direction of the arrow is always clockwise. Even though the angles are measured differently from those in trigonometry, the angle measures are still the same size — just rotated a bit. An angle of 120 degrees is still bigger than a right angle. When you're familiar with the angle sizes, translating into this bearing business is easy.

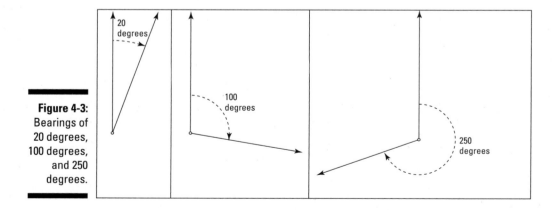

Figure 4-3:
Bearings of
20 degrees,
100 degrees,
and 250
degrees.

Columbus the wizard

It's a given that trigonometry played a big part in navigation and allowed Columbus to find the New World. But trigonometry also helped him in another way. On his voyages, Columbus carried a copy of an almanac created by a mathematician–astronomer by the name of Johannes Müller. In the almanac were tables giving the relative positions of the sun and moon and which determined when and where eclipses would occur. Columbus read that a total eclipse of the moon would occur on February 29, 1504. He took advantage of this information and used it to frighten the natives in the New World into supplying provisions for his ships.

Now, take a look at Figure 4-4 to see the path of a helicopter pilot who flew for 10.5 minutes at a bearing of 36 degrees (which is northeast), then for 13.6 minutes at a bearing of 144 degrees (which is southeast), and then got back to where she started by flying for 14.4 minutes at a bearing of 280 degrees (which is west-northwest).

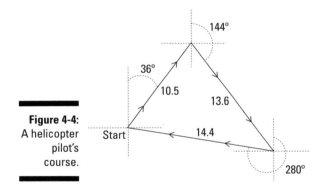

Figure 4-4:
A helicopter pilot's course.

Understanding Norm's Workshop

For those who aren't followers of public television and *The New Yankee Workshop* starring Norm Abram, Norm is a New Englander who does wood-working projects with very expensive tools and invites his audience to do the same (with not-so-expensive tools). He sets his table saw so it can cut a board straight across at a 90-degree angle, or he changes it to cut at any other angle. If Norm wants two perpendicular pieces of wood to meet and form a right angle, he sets his saw at 45 degrees. Figure 4-5 shows how the two pieces of wood fit together.

If Norm wants to create an octagonal (eight-sided) table from a single piece of wood he cuts into eight pieces, then what angles should he cut? More on that in a minute. In Figure 4-6, you see an octagonal table constructed of eight equal triangles.

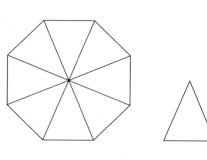

To make his octagonal table, Norm needs eight *isosceles* triangles, where the two long sides are the same length. What are the measures of the angles he has to cut? All the way around a circle (and around the middle of the table) is 360 degrees, so each triangle has a *top angle* (the angle at the center of the table) that measures $\frac{360}{8}$, which is 45 degrees. The two *base angles* (those at the outer edge of the table) are equal in measure. The sum of the measures of the angles of a triangle is 180 degrees, so after subtracting the top angle's 45 degrees, you get 135 degrees for the other two angles together. Dividing the 135 by 2, you find that the base angles are each 67½ degrees. Norm can cut all eight triangles from a single piece of wood, because two base angles plus a top angle form a straight line. He'll just put the triangles together differently after cutting them out. And as you can see from Figure 4-7, he doesn't have much waste.

Figure 4-7:
Cutting eight identical triangles out of a board.

Graphing Angles in Standard Position

Navigators, surveyors, and carpenters all use the same angle measures, but the angles start out in different positions or places. In trigonometry and most other math topics, you draw angles in a standard, universal position, so that mathematicians around the world are drawing and talking about the same thing.

Positioning initial and terminal sides

An angle in *standard position* has its vertex at the origin of the coordinate plane, as shown in Figure 4-8. Its *initial* ray (starting side) lies along the positive *x*-axis. Its *terminal* ray (ending side) moves counterclockwise from the initial side.

If the terminal ray moves clockwise instead of counterclockwise, then the measure is a negative value. You often name this type of angle with a Greek letter.

The lengths of the rays that create the angle have nothing to do with the angle size. You can extend rays as long as you need them to be, and the angle measure won't change. Only the direction of the terminal ray determines the angle.

Measuring by quadrants

Angles in the standard position are used in calculus, geometry, trigonometry, and other math subjects as a basis for discussion. Being able to recognize a particular angle by the quadrant its terminal side lies in and, conversely, to know which angles have their terminal sides in a particular quadrant is helpful.

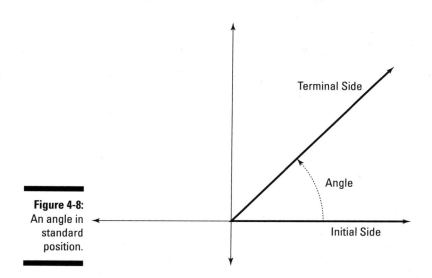

Figure 4-8:
An angle in
standard
position.

Check out Figure 4-9. Angles in standard position that measure between 0 and 90 degrees have their terminal sides in Quadrant I. The angles measuring between 90 and 180 degrees have their terminal sides in Quadrant II. Angles measuring between 180 and 270 have their terminal sides in Quadrant III, and those measuring between 270 and 360 have their terminal sides in Quadrant IV.

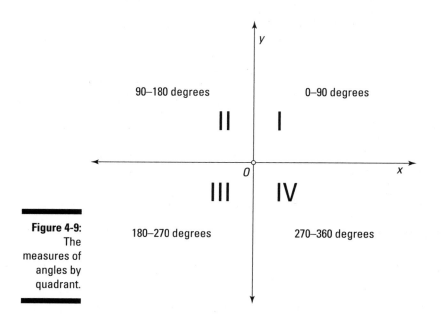

Figure 4-9:
The
measures of
angles by
quadrant.

What's Your Angle? Labeling in Various Ways

The terminal side of an angle determines its angle measure. But more than one angle has the same terminal side — in fact, an infinite number of angles share common terminal sides.

Using negative angle measures

If you want your measurement to be positive, you measure angles in standard position in a *counterclockwise* direction. However, angles can have negative values, too, as you see in Figure 4-10. You get a negative value when you measure angles in a *clockwise* direction. Therefore, an angle of 300 degrees has the same terminal side as an angle measuring –60 degrees. If they have the same terminal side, then why don't they have the same name/size? And which name is better? Sometimes you may want to keep the numerical part of the measure smaller. For example, picturing an angle of –30 degrees is easier than picturing one of 330 degrees. Also, pilots don't always have the choice as to which direction they can turn in, but going 10 degrees in the negative direction makes more sense than going 350 degrees — all the way around, practically — in the positive direction. One common practice is to name all angles with a number that has an absolute value less than 180 degrees. So –60 degrees is often preferable to 300 degrees.

Figure 4-10: Angles with both positive and negative measures.

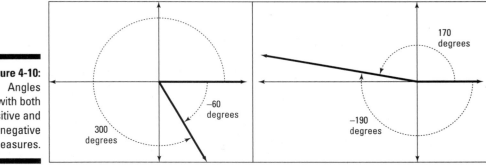

Co-mingling with coterminal angles

Two angles are *coterminal* if they have the same terminal side. You have an infinite number of ways to give an angle measure for a particular terminal ray. Sometimes, using a negative angle rather than a positive angle is more convenient, or the answer to an application may involve more than one revolution (spinning around and around). Angles can have terminal sides that involve one or more full revolutions around the origin or terminal sides that go clockwise instead of counterclockwise — or both of these situations can happen.

More than one revolution

An angle measuring 70 degrees is coterminal with an angle measuring 430 degrees (see Figure 4-11). The angle measuring 430 degrees is actually 360 + 70 (one full revolution plus the original 70). These two angles are also coterminal with an angle of 790 degrees (360 + 360 + 70 = 790). This pattern could go on and on, with the addition of another 360 degrees each time.

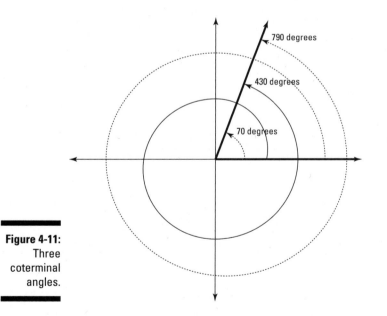

Figure 4-11:
Three
coterminal
angles.

Negative coterminal angles

An angle of 70 degrees is coterminal with an angle of –290 degrees. Two rotations in the negative (clockwise) direction give you an angle of –650 degrees (–290 – 360 = –650).

Renaming angles: So many aliases

Any angle can have many, many descriptions in terms of angle measures, because an angle is equivalent to its coterminal angles. The most frequently used positive angle measures are those that measure between 0 and 360 degrees. Rules for coterminal angles involve adding or subtracting *rotations* (or multiples of 360 degrees). The first equation that follows shows what happens when you add a full rotation over and over. The second shows what happens when you subtract a full rotation many times. The results are all coterminal angles.

$$\theta \to \theta + 360° \to \theta + 720° \to \theta + 1080° \to \dots$$
$$\theta \to \theta - 360° \to \theta - 720° \to \theta - 1080° \to \dots$$

So an angle measuring 100 degrees is coterminal with the following:

Adding: $100° \to 100° + 360° \to 100° + 720° \to 100° + 1080° \to \dots$
                  $100° \to 460° \to 820° \to 1180°$

Subtracting: $100° \to 100° - 360° \to 100° - 720° \to 100° - 1080° \to \dots$
                  $100° \to -260° \to -620° \to -980°$

Here's an example: Suppose you want to give new measures for angles of 800 degrees and −1040 degrees by using an angle measure between 0 and 360 degrees.

1. **Subtract 360 degrees from 800 until the result is less than 360.**

   $800° \to 800° - 360° = 440°$

   $440° \to 440° - 360° = 80°$

   An angle measuring 800 degrees is coterminal with an angle of 80 degrees.

2. **Add 360 degrees to −1040 until the result is positive.**

   $-1040° \to -1040° + 360° = -680°$

   $-680° \to -680° + 360° = -320°$

   $-320° \to -320° + 360° = 40°$

   An angle measuring −1040 degrees is coterminal with an angle of 40 degrees.

How about another example? Give new measures for angles of 280 degrees and −300 degrees by using coterminal angle measures between −180 degrees and 180 degrees.

1. **Subtract 360 from 280.**

   $280° \rightarrow 280° - 360° = -80°$

   An angle measuring 280 degrees is coterminal with an angle of –80 degrees.

2. **Add 360 to –300.**

   $-300° \rightarrow -300° + 360° = 60°$

   An angle measuring –300 degrees is coterminal with an angle of 60 degrees.

The angle measure that you want depends on what form you want the answer to be in. Sometimes, you only want a positive angle measure. And if you're a pilot, you certainly don't want to be given an angle or heading of 2000 degrees. You'd be spinning around until you were dizzy!

Chapter 5

Dishing Out the Pi: Radians

· ·

· ·

A person's first introduction to angles is usually in terms of degrees. You probably have an idea of what a 30-degree angle looks like. (If not, review Chapter 4.) And even most middle-school students know that a triangle consists of 180 degrees. But most of the scientific community uses radians to measure angles. Why change to radians? Why fix what ain't broke? Read on.

What's in a Radian?

A *radian* is much bigger than a degree. Early mathematicians decided on the size of a degree based on divisions of a full circle. A degree is the same as a slice of 1/360 of a circle. No one knows for sure how the choice of 360 degrees in a circle came to be adopted. In any case, 360 is a wonderful number, because you can divide it evenly by so many other numbers: 2, 3, 4, 5, 6, 8, 9, 10, 12, 15, 18, 24, 30, 36, 40, 45, 60, 72, 90, 120, 180, and 360. The early measures of time and distance relied on having convenient numbers to work with. A radian, on the other hand, isn't quite as nice as far as being a whole number. Radians probably were developed because mathematicians wanted to relate the angle measure more to the radius or size of the circle. A circle has 2π radians (a little more than six radians). A radian is *almost* 1/6 of a circle — it's a little more than 57 degrees. Figure 5-1 compares a degree with a radian.

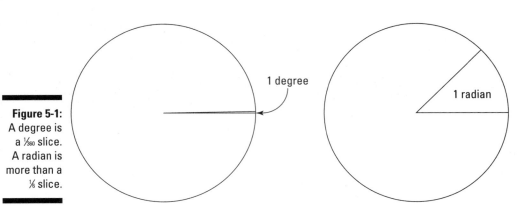

Figure 5-1:
A degree is
a 1/360 slice.
A radian is
more than a
1/6 slice.

1 degree

1 radian

Relating to a circle

The big advantage of using radians is that they're the *natural* measure for dividing up circles. If you took the radius of a circle and bent it into an arc that lies along the circle, you would need a little more than six of them to go all the way around the circle. This fact is true of all circles. An angle measuring one radian would go through the two ends of one of those radii on the circle. The circumference of any circle is always a little more than three times the diameter of that circle — π times the diameter, to be exact. Another way of saying this is 2π times the radius. That number may seem nicer and more civilized than the big number 360, but the disadvantage is that π doesn't have an exact decimal value. Saying 2π radians (which is equal to 360 degrees) means that each circle has *about* 6.28 radians. Even though radians are the natural measure and always relate to the radius and diameter, the decimal values get a bit messy.

Each of these measures has its own place. Measuring angles in degrees is easier, but measuring angles in radians is preferable when doing computations. The radian is more exact because the radius, circumference, or area of the circle is involved. Even though π doesn't have an exact decimal value, when you use multiples of π in answers, they're exactly right. I show you an example of using π as part of an answer in "Making a Clone of Arc," later in this chapter.

Converting degrees and radians

Many math problems require changing from degrees to radian measures or vice versa. You often perform mathematical computations in radians, but then convert the final answers to degrees so the answers are easier to visualize and comprehend. You can use a nifty little proportion to change from degrees to radians or radians to degrees. In this proportion, the Greek letter theta, θ, represents the name of the angle. Putting the superscripts o and R on θ makes the angle stand for the measure in degrees and radians, respectively.

$$\frac{\theta^\circ}{180} = \frac{\theta^R}{\pi}$$

This proportion reads: "The measure of angle θ in degrees divided by 180 is equal to the measure of angle θ in radians divided by π. Remember that π is about 3.141592654.

The computation required for changing degrees to radians and radians to degrees isn't difficult. It has a few tricks, though, and the format is important. You don't usually write the radian measures with decimal values unless you've multiplied through by the decimal equivalent for π.

Changing degrees to radians

To change a measure in degrees to radians, start with the basic proportion for the equivalent angle measures: $\frac{\theta^\circ}{180} = \frac{\theta^R}{\pi}$.

For example, here's how you change a measure of 40 degrees to radians:

1. **Put the 40 in place of the θ° in the proportion.**

   $$\frac{40}{180} = \frac{\theta^R}{\pi}$$

2. **Reduce the fraction on the left.**

   $$\frac{2}{9} = \frac{\theta^R}{\pi}$$

3. **Multiply each side of the proportion by π.**

   $$\pi \cdot \frac{2}{9} = \frac{\theta^R}{\pi} \cdot \pi$$

4. **Simplify the work.**

   $$\frac{2\pi}{9} = \theta^R$$

This example shows that 40 degrees is equivalent to $\frac{2\pi}{9}$ radians. You leave the radian measure as a fraction reduced to lowest terms.

Check out another example: Change a measure of –36 degrees to radians.

1. **Put the –36 in place of the $\theta°$ in the proportion.**

$$\frac{-36}{180} = \frac{\theta^R}{\pi}$$

2. **Reduce the fraction on the left.**

$$-\frac{1}{5} = \frac{\theta^R}{\pi}$$

3. **Multiply each side of the proportion by π.**

$$\pi\left(-\frac{1}{5}\right) = \frac{\theta^R}{\pi} \cdot \pi$$

4. **Simplify the work.**

$$-\frac{\pi}{5} = \theta^R$$

So you see, –36 degrees is equivalent to $-\frac{\pi}{5}$ radians. Having a negative angle is fine (see Chapter 4 for more on negative angles). You leave the expression as a fraction; don't change it to a decimal form.

Changing radians to degrees

You use the same basic proportion to change radians to degrees as you do for changing degrees to radians.

For example, to change $\frac{\pi}{12}$ to a degree measure:

1. **Put the radian measure in place of the $\theta°$ in the proportion.**

$$\frac{\theta°}{180} = \frac{\frac{\pi}{12}}{\pi}$$

2. **Simplify the complex fraction on the right by multiplying the numerator by the reciprocal of the denominator.**

$$\frac{\theta°}{180} = \frac{\pi}{12} \cdot \frac{1}{\pi}$$

$$\frac{\theta°}{180} = \frac{1}{12}$$

3. **Multiply each side of the proportion by 180.**

$$180 \cdot \frac{\theta°}{180} = \frac{1}{12} \cdot 180$$

4. **Reduce and simplify the fraction on the right.**

$$\theta° = \frac{180}{12} = 15°$$

So $\frac{\pi}{12}$ radians is equivalent to 15 degrees.

Just a minute

A full circle contains 360 degrees. If you want just a part of a degree — and a degree is already pretty small — you can say you have ½ of a degree or 0.5 of a degree, or you can use another division. You can divide one degree into 60 *minutes,* and you can divide each minute into 60 *seconds.* So, mathematically, a degree has 3600 subdivisions — you can break it down into 3600 seconds. The way you denote the number of minutes and seconds is with one tick mark (') for minutes and two tick marks (") for seconds. So you read the degree measure 15°45′27″ like this: "Fifteen degrees, 45 minutes, and 27 seconds." Ever since the advent of handheld calculators, people don't use this measure much anymore. The decimal breakdown of a degree is more universally accepted.

Here's another example: Change 1.309 radians to degrees.

I changed this radian measure to a decimal by multiplying through by a decimal equivalent of π, which is approximately 3.1416. You use this same decimal equivalent to solve the problem.

1. **Put the radian measure in place of the $\theta°$ in the proportion.**

$$\frac{\theta°}{180} = \frac{1.309}{\pi}$$

2. **Change the π to a decimal approximation. In this case, I used four decimal places.**

$$\frac{\theta°}{180} = \frac{1.309}{3.1416}$$

3. **Multiply each side of the proportion by 180.**

$$180 \cdot \frac{\theta°}{180} = \frac{1.309}{3.1416} \cdot 180$$

4. **Reduce the fractions, and simplify the value on the right.**

$$\theta° = \frac{5}{12} \cdot 180$$
$$\theta° = 75°$$

This result came out to be a nice number. Sometimes, however, you have a decimal answer for the degrees. Actually, you get a decimal more than sometimes — you *usually* get one.

Highlighting favorites

The favorite or most used angles are those that are multiples of 15 degrees, such as 30, 45, 60, and 90 degrees. Putting these angles into the proportion for changing degrees to radians gives a nice set of angles in radian measure.

First, look at what happens when you replace $\theta°$ with 30:

$$\frac{30}{180} = \frac{\theta^R}{\pi}$$

$$\frac{1}{6} = \frac{\theta^R}{\pi}$$

$$\pi \cdot \frac{1}{6} = \frac{\theta^R}{\pi} \cdot \pi$$

$$\frac{\pi}{6} = \theta^R$$

An angle of 30 degrees is equivalent to $\frac{\pi}{6}$ radians. You get a simple fraction with a π on the top and a nice, small 6 on the bottom.

You get similar results with the other angles:

$$45° \to \frac{\pi}{4},\ 60° \to \frac{\pi}{3},\ 90° \to \frac{\pi}{2}$$

Radian measures with denominators of 2, 3, 4, and 6 are used most frequently.

Making a Clone of Arc

The biggest advantage of using radians instead of degrees is that a radian is directly tied to a length — the length or distance around a circle, which is called its *circumference*. Using radians is very helpful when doing applications involving the length of an *arc* of a circle, which is part of its circumference; measuring the sweep of a hand on a clock; and finding distance in navigation problems.

The problems in this section give you a good sampling of situations where using radians is your best bet. Of course, all these problems presume that you can make *accurate* measurements of the variables you can measure. But trigonometry does open the door to solving practical problems that aren't doable any other way.

Taking chunks out of circles

The examples in this section use parts of circles. A part of a circle may be an arc, a diameter (not really a physical part, but a measure), a *sector* (a piece of the inside), or the center. The measures usually start out in degrees, and I change them to radians, when necessary, to complete the problem.

Scanning with a radar

A radar scans a circular area that has a radius of 40 miles. In one second, it sweeps an arc of 60 degrees. What *area* does the radar cover in one second? In five seconds? Look at Figure 5-2, which shows a sweep of 60 degrees.

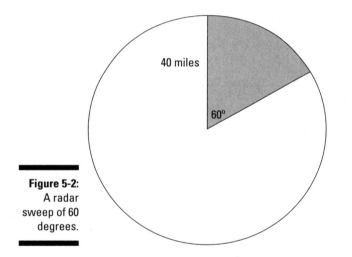

40 miles

60°

Figure 5-2:
A radar sweep of 60 degrees.

Here's how you solve this problem:

1. **Find the area of the circle.**

   Use the formula for the area of a circle, $A = \pi r^2$. Putting the 40 in for the radius, r, you get $\pi r^2 = \pi (40)^2 = 1600\pi \approx 5026.548$ square miles.

2. **Divide by the portion of the circle that the sweep covers.**

   The sweep of 60 degrees is only ⅙ of the entire circle, so you figure the area that the sweep covers by dividing the entire area by 6. The resulting area is $\frac{5026.548}{6} = 837.758$ square miles, which is the area the radar scans in one second. To get the area covered in five seconds, multiply that result by 5 to get 4188.790 square miles.

The preceding problem works out nicely, because the number of degrees is a convenient value — it's a fraction of the circle. But what if the number doesn't divide evenly into 360? For example, what if the radar sweeps an angle of 76 degrees in one second?

In general, if the angle is given in degrees, then the part of a circle that the angle sweeps is $\dfrac{\text{angle in degrees}}{360}$. Take the fraction for that part of the circle and multiply it by the area, πr^2. A fancy name for this part of a circle is *sector*.

Keep the following formulas in mind when you're trying to find the area of a sector:

✔ Using degrees: Area of sector $= \dfrac{\theta^\circ}{360} \cdot \pi r^2$

✔ Or if you're working with radians:

$$\frac{\theta^R}{2\pi} \cdot \pi r^2 = \frac{\theta^R}{2\cancel{\pi}} \cdot \cancel{\pi} r^2 = \frac{\theta^R \cdot r^2}{2} = \text{Area of sector}$$

For example, to find the area of the radar sweep in the preceding example when the radar sweeps 76 degrees per second:

1. **Put 76 in for the θ° and 40 for the radius in the formula for the area of a sector.**

   Area of sweep $= \dfrac{76^\circ}{360} \cdot \pi (40)^2$

2. **Multiply and divide to simplify the answer.**

   Area of sweep $= \dfrac{121,600\pi}{360} \approx \dfrac{382,017.667}{360} = 1,061.160$

To demonstrate this radar-sweep calculation if you're given measurements in radians, find the area of the radar sweep if the sweep is $\frac{\pi}{3}$ radians (which is 60 degrees).

1. **Put $\frac{\pi}{3}$ in for the θ° and 40 in for the radius.**

   Area of sweep $= \dfrac{\theta^R \cdot r^2}{2} = \dfrac{\frac{\pi}{3}(40)^2}{2}$

2. **Multiply and divide to simplify the answer.**

   Area of sweep $= \dfrac{\frac{\pi}{3} \cdot 1600}{2} = \dfrac{1675.516}{2} = 837.758$

Compare this result with the computation for the sweep of 60 degrees, earlier in this section.

Sharing pizza

Some fraternity brothers want to order pizza — and you know how hungry college men can be. The big question is, which has bigger slices of pizza: a 12-inch pizza cut into six slices, or a 15-inch pizza cut into eight slices? Figure 5-3 shows a 12-inch and a 15-inch pizza that are sliced. Can you tell by looking at them which slices are bigger — that is, have more area?

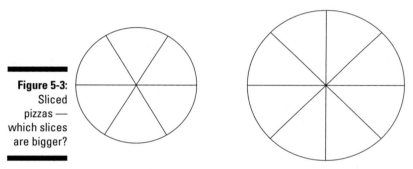

Figure 5-3: Sliced pizzas — which slices are bigger?

The 12-inch pizza is cut into six pieces. Each piece represents an angle of 60 degrees, which is $\frac{\pi}{3}$ radians, so you find the area of each sector (slice) by using the formula for the area of a sector using radians and putting the 6 in for the radius of the pizza with a 12-inch diameter. The answer is

$$\frac{\theta^R \cdot r^2}{2} = \frac{\frac{\pi}{3} \cdot 36}{2} = \frac{12\pi}{2} = 6\pi$$ square inches. (I leave the answer with the multi-

plier of π just so you can compare the sizes between the two pizzas — they'll both have a multiplier of π in them.)

The 15-inch pizza is cut into eight pieces. Each piece represents an angle of 45 degrees, which is $\frac{\pi}{4}$ radians, so, letting the radius be 7.5 this time, the area

of each sector is $\frac{\theta^R \cdot r^2}{2} = \frac{\frac{\pi}{3} \cdot 56.25}{2} = \frac{56.25\pi}{8} = 7.03125\pi$ square inches. This

result doesn't tell you exactly how many square inches are in each slice, but you can see that a slice of this 15-inch pizza has an area of 7.03125π square inches, and a slice of the 12-inch pizza has an area of 6π square inches. The 15-inch pizza has bigger pieces, even though you cut it into more pieces than the 12-inch pizza.

Sweeping hands

I discuss two scenarios in this section: the minute hand of a clock sweeping across the clock's face and the hand of a rider on a Ferris wheel as it whooshes through the air. These examples use the formula for *arc length,* which is the distance around part of a circle.

You find the length of an arc of a circle, *s,* by using the following formula, where the measure of the angle is in radians, and *r* stands for the radius of the circle: $s = r\theta^R$.

Riding the minute hand

Suppose a ladybug settled onto the tip of a tower clock's minute hand. The minute hand is 12 feet long. How far does the ladybug travel from 3:00 until 3:20?

1. **Calculate how many degrees the minute hand swings in 20 minutes.**

    Twenty minutes is ²⁰⁄₆₀ or ⅓ of an hour. Translate that fraction into degrees, and you get ⅓ of 360, or 120 degrees.

2. **Convert degrees to radians.**

    The formula for arc length uses angles in terms of radians, so you first need to change 120 degrees to radians. Using the proportion for changing from degrees to radians and reducing the fraction on the left, $\frac{120°}{180} = \frac{\theta^R}{\pi}, \frac{2}{3} = \frac{\theta^R}{\pi}$, multiply each side of the equation by π. The final result for the angle measure is $\theta^R = \frac{2\pi}{3}$.

3. **Calculate the answer by using the formula for the length of the arc.**

    Enter the angle in radians, and enter 12 (feet), which is the length of the minute hand, for the radius. Your computations should look like $s = \theta^R \cdot r = \frac{2\pi}{3} \cdot 12 = 8\pi$, which is the distance the ladybug traveled, about 25.13 feet.

Riding the Ferris wheel

I don't usually care for Ferris wheels and heights, but on the rare occasions when someone *does* talk me into riding one, I have to face my least favorite part of the whole event: coming down the front of the wheel (I call this part the *front* of the wheel because you can't see the ground below you). Take a look at Figure 5-4. If a Ferris wheel has a 60-foot diameter, how far do you travel from the top of the wheel, halfway down to the bottom?

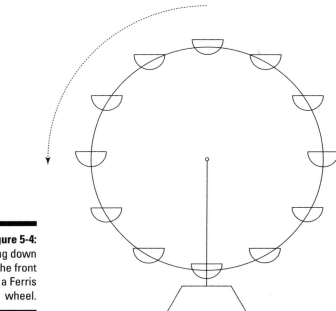

Figure 5-4:
Riding down
the front
of a Ferris
wheel.

A circle with a 60-foot diameter has a 30-foot radius. From the top to the front of the wheel is ¼ of the circle, which is 90 degrees. In radians, 90 degrees is $\frac{\pi}{2}$ (see the section "Highlighting favorites," earlier in this chapter). Using the formula for arc length and putting in the radian measure and the radius of 30 feet, the distance is $s = \theta^R \cdot r = \frac{\pi}{2} \cdot 30 = 15\pi$, which is about 47.12 feet.

Going out and about

One of trigonometry's great qualities is that it lets you measure things that you can't get at or, in the case of the racetrack example in this section, things that you don't want to get close to. A circle and its angles have all sorts of applications both on earth and above.

Measuring the distance to the moon

One of the earliest applications of trigonometry was in measuring distances that you couldn't reach, such as distances to planets or the moon or to places on the other side of the world. Consider the following example.

The diameter of the moon is about 2160 miles. When the moon is full, a person sighting the moon from the earth measures an angle of 0.56 degrees from one side of the moon to the other (see Figure 5-5).

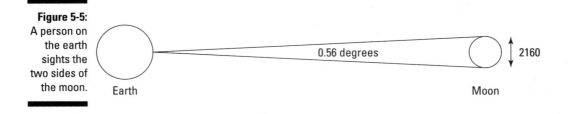

0.56 degrees

2160

Earth

Moon

To figure out how far away the moon is from the earth, consider a circle with the earth at the center and the circumference running right through the center of the moon, along one of the moon's diameters. The moon is so far away that the straight diameter and slight curve of this big circle's circumference are essentially the same measure. The arc that runs through the moon's diameter has an angle of 0.56 degrees and an arc length of 2160 miles (the diameter). Using the arc-length formula, solve for the radius of the large circle, because the radius is the distance to the moon. To solve for the radius:

1. **First, change 0.56 degrees to radians.**

   $$\frac{0.56}{180} = \frac{\theta^R}{\pi}, \theta^R = \frac{0.56\pi}{180} \approx 0.00977$$

2. **Input the numbers into the arc-length formula, $s = \theta^R \cdot r$.**

   Enter 0.00977 radians for the radian measure and 2160 for the arc length: $2160 = 0.00977 \cdot r$.

3. **Divide each side by 0.00977.**

   The distance to the moon is $r = \frac{2160}{0.00977} \approx 221,085$ miles.

Racing around the track

A race car is going around a circular track. A photographer standing at the center of the track takes a picture, turns 80 degrees, and then takes another picture 10 seconds later. If the track has a diameter of ½ mile, how fast is the race car going? Figure 5-6 shows the photographer in the middle and the car in the two different positions.

How fast is the car going? Where does the problem make any mention of speed? Actually, in this situation, the car travels partway around the track in 10 seconds. If you can find out how far the car goes in that time, then you can figure out the car's speed by using the distance formula, where distance is equal to rate multiplied by time.

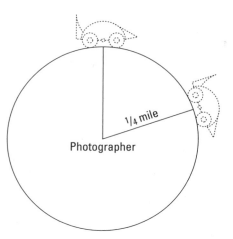

1/4 mile

Photographer

Figure 5-6:
A car races
around the
track.

A formula that you'll find mighty helpful is the one that says distance equals rate multiplied by time, where rate is miles per hour (or feet per second or some such measure), and time is the same measure as in the rate:

$$d = r \cdot t.$$

1. **First, change the 80 degrees to radians.**

   You end up with $\frac{4\pi}{9}$ radians.

2. **Input the numbers in the arc-length formula.**

   Putting in the radian measure and the radius of the track (¼ mile), you get arc length $= \frac{4\pi}{9} \cdot \frac{1}{4} = \frac{\pi}{9} \approx 0.349$ mile, which is the distance the car traveled in 10 seconds.

3. **Multiply this result by 6 to get miles traveled in a minute.**

   This calculation gives you 2.094 miles per minute.

4. **Then multiply that number by 60 to get miles traveled in one hour.**

   This calculation gives you 125.64 miles. So the car is traveling about 125 mph.

Chapter 6

Getting It Right with Triangles

*T*riangles are classified in many ways. One way of distinguishing one triangle from another is to use angle measurements. Because a 90-degree angle is called a *right angle,* you use the same terminology for a triangle with a right angle in it. This type of triangle is called a *right triangle*. And that's all right.

The measures of the sides of right triangles determine the values of the trig functions. And those trig functions (along with the right triangles) are really handy when it comes to solving problems such as "Just how high is that tree?" The special properties of a right triangle — some credited with Pythagoras — make them very useful in trigonometry and other math areas.

Sizing Up Right Triangles

If you're looking at their angles, triangles can be right, acute, or obtuse. Right triangles have been of great interest for centuries. They're the basis for studies in navigation, astronomy, surveying, and military engineering.

What's so right about them?

A *right triangle* has a right angle in it. But it can only have one right angle, because the total number of degrees in a triangle is 180. If it had two right angles, then those two angles would take up all 180 degrees; no degrees would be left for a third angle. So in a right triangle, the other two angles share the remaining 90 degrees.

Right triangles can come in all sorts of shapes, but they all have that corner, where the right angle sits. In Figure 6-1, you see that in all the triangles, the right angle has two sides that are perpendicular to one another. The other two angles are *acute* angles (meaning they're less than 90 degrees).

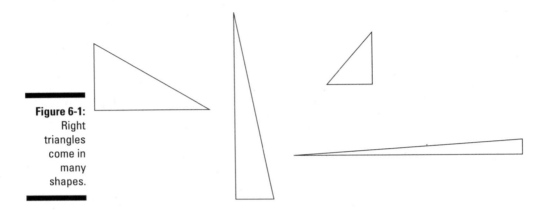

Figure 6-1:
Right triangles come in many shapes.

The anatomy of a right triangle

Right triangles are a familiar sight — not just in geometry classes. Carpenters have tools for measurement that are right triangles. Architects who design by hand (rather than on a computer) draw with right-triangle templates. Even though the focus in a right triangle is the right angle, a right triangle actually has six different parts: three angles and three sides. Now, this fact is true of any triangle, but right triangles have special names for these parts. Having special names is necessary because so many properties, theorems, and applications using right triangles are out there, and the names make talking about and explaining the triangles more clear.

Figure 6-2 shows a typical right triangle labeled with capital letters and lower-case letters. Since the time of Leonard Euler, the famous Swiss mathematician, this type of labeling has been the tradition. You use lowercase letters to mark the sides of the triangle and capital letters to mark the *vertex* (angle) opposite the side with the corresponding lowercase letter.

The little square at the vertex C shows that the two sides meeting there are perpendicular at that vertex — that's where the right angle is. The side c, opposite the right angle, is called the *hypotenuse*. The other two sides, a and b, are called the *legs*. The hypotenuse is always the longest side, because it's opposite the largest angle.

Squaring the corners

Over 20 years ago, my spouse and I selected house plans that we liked and started all the processes needed to build our new house. We didn't have all that much money, so we did as much as we could by ourselves. After the lot was cleared of bushes and weeds, we went there with pegs and string to lay out the foundation. The backhoe was due the next day to dig the hole for the basement. With blueprints in hand, we had all the measurements — how long each side was to be. A tape measure gave us accurate measures for lengths, but what about the right angles? A school protractor isn't big enough to make those corners square. We used a method called *squaring the corners*. We knew that some nice measures for the three sides of a right triangle are 3, 4, 5 or 6, 8, 10. At a corner in question, we would pick a point on one string that was 4 feet from the peg and a point on the other string that was 3 feet from the peg. Then we'd measure diagonally from each of those points. If the diagonal measure didn't come out to be 5 feet, then we didn't have a right angle — the corner wasn't square. It took a lot of peg moving, but we got the foundation laid out accurately. The house is still standing — in fact, we still live in it — so I guess we did a good job.

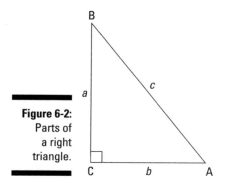

Figure 6-2:
Parts of a right triangle.

In Figure 6-2, the angle at vertex *C* is the right angle, and the other two angles, *A* and *B*, are acute angles. If the measures of the angles at *A* and *B* are the same, then they're each 45 degrees, and the triangle is *isosceles*. If that's the case, then the lengths of the sides *a* and *b* are the same also.

If the angle at vertex *A* is bigger than that at vertex *B*, then side *a* is longer than side *b*. The measures of a triangle's angles have a direct relationship to the lengths of the sides opposite them.

Complementary angles

Complementary angles don't say to one another, "Oh, how acute you are!" This term indicates that the measures of two angles add up to 90 degrees. Indicating whether two angles are complementary may not seem to be a big deal, but having complementary angles can create some interesting situations when you use their function values. Find out more in Chapters 10 and 11.

In a right triangle, the biggest angle is always 90 degrees. So the remaining two angles have 90 degrees to share, which means that they're complementary angles. If angle *A* measures 25 degrees, then angle *B* must be 65 degrees, because $90 - 25 = 65$.

Pythagoras Schmythagoras: Demystifying the Pythagorean Theorem

Pythagoras was a Greek mathematician who lived around 570 B.C. Even with the relatively primitive tools at his disposal, he was able to discover and formulate a *theorem,* or rule, that became one of the most well-known in all of mathematics: the Pythagorean theorem.

The *Pythagorean theorem* says that if *a, b,* and *c* are the sides of a right triangle, as shown in Figure 6-3, and if *c* is the side opposite the right angle, then their lengths have the following property: $a^2 + b^2 = c^2$.

Figure 6-3:
The sides of this right triangle are labeled as in the Pythagorean theorem.

Have you heard the one about the squaw on the hippopotamus?

The first time I heard this joke was on a Chicago radio station in the 1960s. The disc jockey was Howard Miller — as many Chicagoans will agree, a man before his time. Who else could get away with telling a math joke on morning radio? The joke goes as follows: There was an Indian tribe with the usual hierarchy of braves, chiefs, and squaws. Two of the braves had squaws who were expecting papooses. The chief's squaw was also expecting a papoose.

The squaws of the braves sat working on buffalo hides, grinding corn and doing what squaws do. But the squaw of the chief sat on a hippopotamus hide to do her daily chores. The happy days arrived; the papooses were born. The squaws of the two braves each had sons. And the squaw of the chief had *twin* sons. The moral of this story is "The squaw on the hippopotamus is equal to the sons of the squaws on the other two hides." Yes, I'm sure you're groaning.

You can work backwards with the Pythagorean theorem, too. For example, if you don't know what type of triangle you have, and the sides of a triangle are 3, 4, and 5 units in length, then the triangle must be a right triangle, because

$$3^2 + 4^2 = 5^2$$
$$9 + 16 = 25$$

The sum of the squares of the two shorter sides, which are called the *legs*, is equal to the square of the longest side, which is called the *hypotenuse*.

Hitting a Pythagorean triple

A *Pythagorean triple* is a list of three numbers that fits the Pythagorean theorem — the square of the largest number is equal to the sum of the squares of the two smaller numbers. The multiple of any Pythagorean triple (multiply each of the numbers in the triple by the same number) is also a Pythagorean triple. They seem to reinvent themselves.

Familiarizing yourself with the more frequently used Pythagorean triples is very helpful. If you recognize that you have a triple, then working with applications is much easier.

Table 6-1 shows some of the most common Pythagorean triples and some of their multiples.

Table 6-1	Common Pythagorean Triples		
Triple	*Triple × 2*	*Triple × 3*	*Triple × 4*
3, 4, 5	6, 8, 10	9, 12, 15	12, 16, 20
5, 12, 13	10, 24, 26	15, 36, 39	20, 48, 52
7, 24, 25	14, 48, 50	21, 72, 75	28, 96, 100
9, 40, 41	18, 80, 82	27, 120, 123	36, 160, 164
11, 60, 61	22, 120, 122	33, 180, 183	44, 240, 244

Here's how to check out a triple by using the Pythagorean theorem. Try out the triple 9, 40, 41:

1. **Replace *a*, *b*, and *c* with 9, 40, and 41, respectively.**

   $$9^2 + 40^2 = 41^2$$
   $$81 + 1600 = 1681$$

2. **Then replace *a*, *b*, and *c* with the 9-40-41 triple multiplied by 3 (27, 120, 123).**

   $$27^2 + 120^2 = 123^2$$
   $$729 + 14,400 = 15,129$$

Solving for a missing length

One of the nice qualities of right triangles is the fact that you can find the length of one side if you know the lengths of the other two sides. You don't have this luxury with just any triangle, so count your blessings now.

Practicing on triangles

The Pythagorean theorem states that $a^2 + b^2 = c^2$ in a right triangle where c is the longest side. You can use this equation to figure out the length of one side. Figure 6-4 shows two right triangles that are each missing one side's measure.

In the left triangle in Figure 6-4, the measure of the hypotenuse is missing. Use the Pythagorean theorem to solve for the missing length.

1. **Replace the variables in the theorem with the values of the known sides.**

   $$48^2 + 14^2 = c^2$$

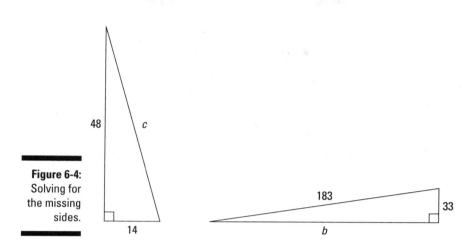

Figure 6-4:
Solving for
the missing
sides.

2. **Square the measures and add them together.**

$$2304 + 196 = c^2$$
$$2500 = c^2$$

3. **Find the square root of the sum.**

$$\sqrt{2500} = \sqrt{c^2}$$
$$50 = c$$

The length of the missing side, c, which is the hypotenuse, is 50.

The triangle on the right in Figure 6-4 is missing the bottom length, but you do have the length of the hypotenuse. It doesn't matter whether you call the missing length a or b.

1. **Replace the variables in the theorem with the values of the known sides.**

$$33^2 + b^2 = 183^2$$

2. **Square the measures, and subtract 1089 from each side.**

$$1089 + b^2 = 33,489$$
$$b^2 = 32,400$$

3. **Find the square root of the difference.**

$$\sqrt{b^2} = \sqrt{32,400}$$
$$b = 180$$

The length of the missing side is 180 units. That's not much shorter than the hypotenuse, but it still shows that the hypotenuse has the longest measure.

Finding the distance across a pond

Trigonometry is very handy for finding distances that you can't reach to measure. Imagine that you want to string a cable diagonally across a pond (so you can attach a bunch of fishing line and hooks). The diagonal distance is the hypotenuse of a right triangle. You can measure the other two sides along the shore. Figure 6-5 shows the pond and the imaginary right triangle you use to figure out how long your cable needs to be.

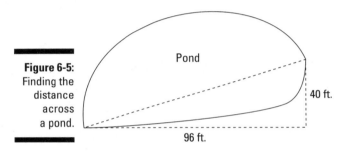

Figure 6-5:
Finding the distance across a pond.

Pond

40 ft.

96 ft.

The two sides of the triangle that you can measure, the height and the width of the pond, are 40 feet and 96 feet. These are the two legs of a right triangle. Use the Pythagorean theorem to solve for the hypotenuse, which is the diagonal distance across the pond.

1. **Replace the variables in the theorem with the values of the known sides.**

   $40^2 + 96^2 = c^2$

2. **Square the measures, and add them together.**

   $1600 + 9216 = c^2$

   $10,816 = c^2$

3. **Find the square root of the sum.**

   $\sqrt{10,816} = \sqrt{c^2}$

   $104 = c$

The diagonal across the pond is 104 feet. String up your cable, and go fishing!

In a League of Their Own: Special Right Triangles

Right triangles are handy little suckers. The relationship between the lengths of the sides helps you measure lengths that you can't reach. And just when you think that math can't get any better, along come two triangles that are the cat's meow. One of them is an isosceles right triangle. The two legs are the same, and the hypotenuse is always a multiple of their length. The other special right triangle has one side half as long as the other. These two triangles are very useful, because the angle measures in them are some of the most popular.

30-60-90 right triangles

A 30-60-90 right triangle has angles measuring just what the name says. The two acute, complementary angles are 30 and 60 degrees. These triangles are great to work with, because the angle measures, all being multiples of 30, have a pattern, and so do the measures of the sides. Oh, yes, the Pythagorean theorem still holds — you have that relationship between the squares of the sides. But a, b, and c are related in another way, too. In a 30-60-90 right triangle, if a is the shortest side, then the hypotenuse, the longest side, measures twice that, or $2a$. You can use $2a$ instead of the letter c. And the middle length is $\sqrt{3}\,a$, or about 1.7 times as long as the shortest side; this number replaces the letter b. The particularly nice part about this triangle is that you can write all three sides in terms of one variable, a. Look at how these lengths fit into the Pythagorean theorem:

$$a^2 + \left(\sqrt{3}\,a\right)^2 = (2a)^2$$
$$a^2 + 3a^2 = 4a^2$$

Figure 6-6 shows the 30-60-90 right triangle.

Here's a sample problem you can solve by taking advantage of the special relationships within a 30-60-90 right triangle: If the hypotenuse of a 30-60-90 right triangle is 8 units long, then how long are the other two sides?

1. **Find the length of the shorter leg.**

   The hypotenuse is twice as long as the shorter leg, a. So 8 = $2a$. Divide by 2, and you get a = 4.

2. **Find the length of the longer leg.**

   The longer leg is $\sqrt{3}\,a$, so multiply $\sqrt{3}$ times 4 to get $4\sqrt{3}$, or about 6.9 units.

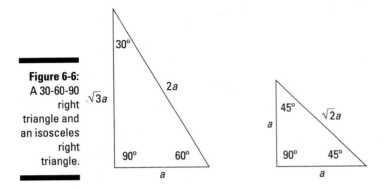

Figure 6-6:
A 30-60-90
right
triangle and
an isosceles
right
triangle.

Isosceles right triangles

The other special right triangle is the *isosceles right triangle,* or the 45-45-90 right triangle. The two acute angles are equal, making the two legs opposite them equal, too. What's more, the lengths of those two legs have a special relationship with the hypotenuse (in addition to the one in the Pythagorean theorem, of course). In an isosceles right triangle, if the legs are each a units in length, then the hypotenuse is $\sqrt{2}\,a$, or about 1.4 times as long as a leg. Figure 6-6 shows a sample isosceles right triangle.

Now that you know how isosceles right triangles work, try your hand at this sample problem: If an isosceles right triangle has a hypotenuse that's 16 units long, then how long are the legs?

1. **Create an equation to solve.**

   The hypotenuse is $\sqrt{2}\,a$, where a is the length of the legs. You know that the hypotenuse is 16, so you can solve the equation $\sqrt{2}\,a = 16$ for the length of a.

2. **Solve for a.**

   Divide each side by the radical to get

   $$\frac{\sqrt{2}\,a}{\sqrt{2}} = \frac{16}{\sqrt{2}}$$

   $$a = \frac{16}{\sqrt{2}} = \frac{16}{\sqrt{2}} \cdot \frac{\sqrt{2}}{\sqrt{2}} = \frac{16\sqrt{2}}{2} = 8\sqrt{2}$$

   Each leg is about 11.3 units.

Part II
Trigonometric Functions

The 5th Wave   By Rich Tennant

"IT SAYS HERE IF I SUBSCRIBE TO THIS MAGAZINE, THEY'LL SEND ME A FREE DESK-
TOP CALCULATOR. DESKTOP CALCULATOR?!! WHOOAA – WHERE HAVE I BEEN?!!"

In this part . . .

You start with the right triangle and the definition of the six trig functions using the sides of that triangle. Then your horizons expand as you discover how to use these same trig functions with a circle. Just to keep you grounded, I provide some down-to-earth trig applications.

Chapter 7

Doing Right by Trig Functions

· ·

In This Chapter

▶ Understanding the three basic trig functions

▶ Building on the basics: The reciprocal functions

▶ Recognizing the angles that give the cleanest trig results

▶ Determining the exact values of functions

· ·

*B*y taking the lengths of the sides of right triangles or the chords of circles and creating ratios with those numbers and variables, our ancestors marked the birth of trigonometric functions. These functions proved to be of infinite value, as they allow you to use the stars to navigate and to build bridges that won't fall. If you're not into navigating or engineering, then you can use the trig functions at home to plan that new addition. They're a staple for students going into calculus.

The six trig functions require one thing of you — inputting an angle measure — and then they output some numbers. These outputs can be any real numbers, from infinitely small to infinitely large and everything in between. The results you get depend on which function you use. Although some of the earlier computations were rather tedious, today's hand-held calculators make everything much easier.

SohCahToa to the Rescue: How Trig Functions Work

What or who is *SohCahToa* — an Italian pasta dish, an American Indian princess, or some new miracle drug, perhaps? Actually, none of the above. Some clever math teacher made up this word in order to help students remember the trig ratios. The word then got around. Before I explain *SohCahToa*, you need to know what the letters stand for. You can see the letters *S, C,* and *T* for the trig functions sine, cosine, and tangent. The lowercase letters represent their ratios in a right triangle.

The name game: A right triangle's three sides

A right triangle has two shorter sides, or *legs,* and the longest side, opposite the right angle, which is always called the *hypotenuse.* The two shorter sides have some other special names, too, based on which acute angle of the triangle you happen to be looking at.

In reference to acute angle θ (see Figure 7-1), the leg on the other side of the triangle from θ is called the *opposite* side. That opposite side isn't along one of the rays making up the angle. The other leg in a right triangle is then called the *adjacent* side. Adjacent means "next to," and in the case of right triangles, the adjacent side helps form the acute angle along with the hypotenuse because it lies along one of the angle's rays.

Figure 7-1:
The acute angle θ determines the names of the right triangle's sides.

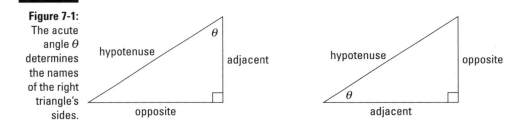

The six ratios: Relating the three sides

Each of the three sides of a right triangle — hypotenuse, opposite, and adjacent — has a respective length or measure. And those three lengths or measures form six different ratios. Check out Figure 7-2, which has sides of lengths 3, 4, and 5.

The six different ratios that you can form with the numbers 3, 4, and 5 are

$$\frac{3}{4}, \frac{3}{5}, \frac{4}{3}, \frac{4}{5}, \frac{5}{3}, \frac{5}{4}$$

How far to the moon?

Hipparchus lived from about 190 to 120 B.C. Also known as Hipparchus of Nicaea, he did his astrological observations from Rhodes between 146 and 127 B.C. He was the first astronomer to compile a catalog of 850 stars, well before telescopes were available, and his computations were remarkably accurate. He obtained measurements of the length of the year and the distance to the moon. To measure the distance to the moon, Hipparchus and a colleague each observed a solar eclipse — a total eclipse at Syene and a partial eclipse at Alexandria, where ⅚ of the sun was blocked. Using angles and the distance between the two cities, thus creating an imaginary triangle with the lines of sight from those two cities to the moon, he was able to make his calculations. The trig ratios he used were on a big scale, but it doesn't matter how large the triangle is, because the trig functions for the angles don't change with the triangle's size. Hipparchus is commemorated by having a moon crater, a Mars crater, and an asteroid named after him.

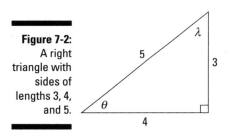

Figure 7-2:
A right triangle with sides of lengths 3, 4, and 5.

These six fractions are all that you can make by using the three lengths of the sides. The ratios are special because they represent all the possible output values of the trig functions for the acute angles in that triangle — in short, you can figure out the value of an unknown angle in a right triangle just by computing a trig function of one of these ratios.

The sine function: Opposite over hypotenuse

When you're using right triangles to define trig functions, the trig function *sine,* abbreviated *sin,* has input values that are angle measures and output values that you obtain from the ratio $\frac{\text{opposite}}{\text{hypotenuse}}$. Figure 7-2 (in the preceding section) shows two different acute angles, and each has a different value for the function sine. The two values are $\sin \theta = \frac{3}{5}$ and $\sin \lambda = \frac{4}{5}$.

The sine is always the measure of the opposite side divided by the measure of the hypotenuse. Because the hypotenuse is always the longest side, the number on the bottom of the ratio will always be larger than that on the top. For this reason, the output of the sine function will always be a proper fraction — it'll never be a number equal to or greater than 1 unless the opposite side is equal in length to the hypotenuse (which only happens when you're working with circles — see Chapter 8).

Even if you don't know both lengths required for the sine function, you can calculate the sine if you know any two of the three lengths of a triangle's sides. For example, to find the sine of angle α in a right triangle whose hypotenuse is 10 inches long and adjacent side is 8 inches long:

1. **Find the length of the side opposite α.**

   Use the Pythagorean theorem, $a^2 + b^2 = c^2$ (see Chapter 6), letting a be 8 and c be 10. When you input the numbers and solve for b, you get

   $$8^2 + b^2 = 10^2$$
   $$64 + b^2 = 100$$
   $$b^2 = 36,\ b = 6$$

   So the opposite side is 6 inches long.

2. **Use the ratio for sine, opposite over hypotenuse.**

   $$\sin \alpha = \frac{\text{opposite}}{\text{hypotenuse}} = \frac{6}{10} = \frac{3}{5}$$

The cosine function: Adjacent over hypotenuse

The trig function *cosine,* abbreviated *cos,* works by forming this ratio: $\frac{\text{adjacent}}{\text{hypotenuse}}$. Take a look back at Figure 7-2, and you see that the cosines of the two angles are $\cos \theta = \frac{4}{5}$ and $\cos \lambda = \frac{3}{5}$. The situation with the ratios is the same as with the sine function — the values are going to be less than or equal to 1 (the latter only when dealing with circles), never greater than 1, because the hypotenuse is the denominator.

The two ratios for the cosine are the same as those for the sine — except the angles are reversed. This property is true of the sines and cosines of *complementary* angles in a right triangle (meaning those angles that add up to 90 degrees).

If θ and λ are the two acute angles of a right triangle, then $\sin \theta = \cos \lambda$ and $\cos \theta = \sin \lambda$.

Now for an example. To find the cosine of angle β in a right triangle if the two legs are each $\sqrt{6}$ feet in length:

1. **Find the length of the hypotenuse.**

   Using the Pythagorean theorem, $a^2 + b^2 = c^2$ (see Chapter 6), and replacing both a and b with the given measure, solve for c.

   $$\left(\sqrt{6}\right)^2 + \left(\sqrt{6}\right)^2 = c^2$$
   $$6 + 6 = c^2$$
   $$12 = c^2, c = \sqrt{12} = 2\sqrt{3}$$

   The hypotenuse is $2\sqrt{3}$ feet long.

2. **Use the ratio for cosine, adjacent over hypotenuse, to find the answer.**

   $$\cos \beta = \frac{\text{adjacent}}{\text{hypotenuse}} = \frac{\sqrt{6}}{2\sqrt{3}} = \frac{\sqrt{2}}{2}$$

The tangent function: Opposite over adjacent

The third trig function, *tangent,* is abbreviated *tan.* This function uses just the measures of the two legs and doesn't use the hypotenuse at all. The tangent is described with this ratio: $\frac{\text{opposite}}{\text{adjacent}}$. No restriction or rule on the respective sizes of these sides exists — the opposite side can be larger, or the adjacent side can be larger. So the tangent ratio produces numbers that are very large, very small, and everything in between. If you hike on back to Figure 7-2, you see that the tangents are $\tan \theta = \frac{3}{4}$ and $\tan \lambda = \frac{4}{3}$. And in case you're wondering whether the two tangents of the acute angles are always *reciprocals* (flips) of one another, the answer is yes. The trig identities in Chapter 10 explain this phenomenon.

The following example shows you how to find the values of the tangent for each of the acute angles in a right triangle where the hypotenuse is 25 inches and one leg is 7 inches.

1. **Find the measure of the missing leg.**

   Using the Pythagorean theorem, $a^2 + b^2 = c^2$ (see Chapter 6), putting in 7 for a and 25 for c, and solving for the missing value, b, you find that the unknown length is 24 inches:

$$7^2 + b^2 = 25^2, \ b^2 = 25^2 - 7^2 = 625 - 49 = 576$$
$$b^2 = 576, \ b = 24$$

2. **Select names for the acute angles in order to determine the *opposite* and *adjacent* designations.**

   The easiest way to do this is to draw a picture and label it — take a look at Figure 7-3.

Figure 7-3:
Labeling
a right
triangle and
naming
the acute
angles.

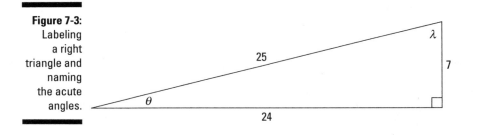

The two acute angles are named with the Greek letters θ and λ. The side opposite θ measures 7 inches, and the side adjacent to it measures 24 inches. For angle λ, the opposite side measures 24 inches, and the adjacent side measures 7 inches.

3. **Form the two tangent ratios by using the values 7, 24, and 25.**

   $$\tan \theta = \frac{\text{opposite}}{\text{adjacent}} = \frac{7}{24}, \ \tan \lambda = \frac{\text{opposite}}{\text{adjacent}} = \frac{24}{7}$$

All together, now: Using one function to solve for another

Sometimes you have to solve for a trig function in terms of another function. In the following example, the cosine of angle λ is $\frac{12}{13}$. What are the values of the sine and tangent of λ?

1. **Identify the sides given by the cosine function.**

   The cosine ratio is $\frac{\text{adjacent}}{\text{hypotenuse}}$. Using the given ratio, the adjacent side measures 12 units, and the hypotenuse measures 13 units.

2. **Find the measures of the missing side.**

   Using the Pythagorean theorem, you find that the missing side (the opposite side) measures 5 units.

$$12^2 + b^2 = 13^2, \ b^2 = 13^2 - 12^2, \ b^2 = 169 - 144 = 25$$
$$b^2 = 25, \ b = 5$$

3. **Determine the values of the sine and tangent.**

The sine is $\dfrac{\text{opposite}}{\text{hypotenuse}}$ and the tangent is $\dfrac{\text{opposite}}{\text{adjacent}}$, so

$\sin \lambda = \dfrac{5}{13}$ and $\tan \lambda = \dfrac{5}{12}$.

Socking the rules away: The legend of SohCahToa

And now, for the fun part: the legend of an Indian chief named SohCahToa (read that: *soak-a-toe-ah*). Many years ago, a tribe of American Indians lived along the Illinois River, where they hunted and fished and did what was necessary to live in peace. One young brave was trying to learn to use his bow and arrows effectively, and he was having all sorts of trouble. Out of frustration, he kicked what he thought was something soft, but it was a rock. His toe turned blue and throbbed all day and night. He tried wrapping it, rubbing it, and ignoring it, but nothing gave him any relief. His mother, a wise squaw, finally had enough of his complaining and said, "Go down to the river, now, and *Soh Cah Toa!*" The young brave went to the river, put his toe in the cool water, and got relief. He never did get proficient with the bow and arrow, and he kept kicking things in frustration. Pretty soon, he was known as SohCahToa.

Sure, this story is pretty lame, but you'll find it very useful when trying to remember the ratios for the three basic trig functions:

✔ **Soh** stands for Sine Opposite (over) Hypotenuse.

✔ **Cah** is Cosine Adjacent (over) Hypotenuse.

✔ **Toa** is Tangent Opposite (over) Adjacent.

People who studied trigonometry in the past may not remember too many details about it, but one thing they *do* remember is SohCahToa, if they've heard the story (or something similar).

Taking It a Step Further: Reciprocal Functions

The three most basic trig functions (sine, cosine, and tangent) use the three sides of a triangle, taking two at a time and making fractions of them. But three more trig functions exist, and these are called the *reciprocal* functions because they use the reciprocals, or flips, of the original three functions. If these last three functions are just reciprocals of the first three, why are they even necessary? You could probably live without them, but you'd miss them when doing calculations and solving equations. Having an unknown variable in the numerator of a fraction when solving an equation is just nicer and more convenient than having one in the denominator, and these reciprocal functions make that situation possible. It's nothing more than convenience, but I'm all for such luxury.

The cosecant function: Sine flipped upside down

The *cosecant* function, abbreviated *csc,* is the reciprocal of the sine function and thus uses this ratio: $\dfrac{\text{hypotenuse}}{\text{opposite}}$. The hypotenuse of a right triangle is always the longest side, so the numerator of this fraction is always larger than the denominator. As a result, the cosecant function always produces values bigger than 1.

You can use the values in Figure 7-4 to determine the cosecants of the two acute angles: $\csc \theta = \dfrac{13}{12}$ and $\csc \lambda = \dfrac{13}{5}$.

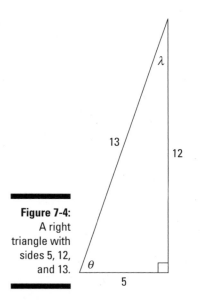

Figure 7-4:
A right triangle with sides 5, 12, and 13.

Suppose someone asks you to find the cosecant of angle α if you know that the hypotenuse is 1 unit long and that the right triangle is isosceles. Remember that an isosceles triangle has two congruent sides (flip back to Chapter 6 for a refresher). These two sides have to be the two legs, because the hypotenuse has to have the longest side. So to find the cosecant:

1. **Find the lengths of the two legs.**

   The Pythagorean theorem says that $a^2 + b^2 = c^2$, but because two sides are congruent, you can take out one variable and write the equation as $a^2 + a^2 = c^2$. Put in 1 for c and solve for a.

   $$2a^2 = 1$$

   $$a^2 = \frac{1}{2}$$

   $$a = \sqrt{\frac{1}{2}} = \frac{1}{\sqrt{2}}$$

   The legs are both $\frac{1}{\sqrt{2}}$ units long. You can leave the radical in the denom-

   inator and not worry about rationalizing, because you're going to input the whole thing into the cosecant ratio, anyway, and things can change.

2. **Use the length of the opposite side in the ratio for cosecant.**

   $$\csc \alpha = \frac{\text{hypotenuse}}{\text{opposite}} = \frac{1}{\frac{1}{\sqrt{2}}} = \sqrt{2}$$

The secant function: Cosine on its head

The *secant* function, abbreviated *sec,* is the reciprocal of the cosine. So its ratio is $\frac{\text{hypotenuse}}{\text{adjacent}}$. Just as with the cosecant, the ratio of the sides is greater than 1. Using the triangle in Figure 7-4, the two secants are $\sec \theta = \frac{13}{5}$ and $\sec \lambda = \frac{13}{12}$.

The cotangent function: Tangent, tails side up

The last reciprocal function is the *cotangent,* abbreviated *cot.* This function is the reciprocal of the tangent (hence, the *co-*). The ratio of the sides for the cotangent is $\frac{\text{adjacent}}{\text{opposite}}$. So if you look back at Figure 7-4, you see that the two cotangents are $\cot \theta = \frac{5}{12}$ and $\cot \lambda = \frac{12}{5}$.

Seeing the sun at the center of it all

Before the 16th century, scientists believed that the earth was the center of our solar system. Nicolaus Copernicus, a Polish monk, introduced the solar system as we know it today — with the sun in the center. He was able to deduce this by using mathematical calculations, including trigonometry, and observations of the stars and planets made with the naked eye. The church didn't like this theory, burning many people to death for defending Copernicus's idea. Convincing everyone of this radical new idea was difficult because telescopes didn't yet exist, and many things weren't visible to the naked eye. Some credit Copernicus with saying, "If the sense of sight could ever be made sufficiently powerful, we could see the phases of the planet, and you would be convinced." Within a century, Galileo invented the telescope and revealed just those phases of the planets that Copernicus referred to. Further evidence came from the work of Johannes Kepler, who determined that the planets move in elliptical paths, not circular, and Isaac Newton, who provided the additional mathematical processes to formalize the theories.

The ratio for the cotangent is just the ratio, not necessarily the lengths of the sides. The fraction made by the lengths might've been reduced by dividing numerator and denominator by the same number.

Sometimes you know the value of the cotangent along with other info and have to solve for one or both of the sides. Try this example: What are the lengths of the legs of a right triangle if $\cot \alpha = \frac{11}{60}$ and the hypotenuse is 183 inches long?

1. **Write the adjacent and opposite sides as multiples of the same number, *m*, and put them in the Pythagorean theorem with the hypotenuse.**

   $$(11m)^2 + (60m)^2 = 183^2$$

2. **Simplify the equation and solve for *m*.**

   $$(11m)^2 + (60m)^2 = 183^2$$
   $$121m^2 + 3600m^2 = 33,489$$
   $$3721m^2 = 33,489$$
   $$m^2 = \frac{33,489}{3721} = 9$$
   $$m = 3$$

3. **Use the value of *m* to find the lengths of the two legs.**

   Because you know that $m = 3$, you know that the adjacent side is $11m = 11 \cdot 3 = 33$ and the opposite side is $60m = 60 \cdot 3 = 180$. The three sides of the right triangle are 33, 180, and 183. You can double-check your results by plugging these three numbers into the Pythagorean theorem and making sure the theorem holds true.

Angling In on Your Favorites

You may have a favorite television show, dessert, or color. Usually, however, a favorite *angle* isn't near the top of the list. But a favorite angle isn't really out of line in the scheme of things. My favorite angle is a 30-degree angle — there's just something so acute about it.

Identifying the most popular angles

The most common or popular angles are those with measures that are multiples of 15 degrees. Topping the list are 30, 60, and 90 degrees. Another favorite is 45 degrees. The reason that these angles are favorites is because they all divide 360 degrees evenly. In Figure 7-5, the circle is divided into 12 slices that are each 30 degrees and 24 slices that are each 15 degrees. These exact divisions result in nicer-than-usual values for the different trig functions of the angles.

Figure 7-5:
A circle divides neatly into 30-degree and 15-degree slices.

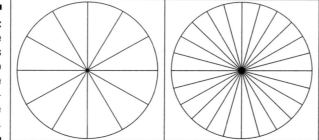

One way to capitalize on the four main angles — 30, 45, 60, and 90 — is to look at their multiples that go up to 360 degrees. The trig functions of the first four basic angles and the trig functions of their multiples are related, which I cover in detail in Chapter 8. The list of all-time favorites includes multiples of 30 degrees (30, 60, 90, 120, 150, 180, 210, 240, 270, 300, and 330) and some multiples of 15 degrees that are between them: 45, 135, 225, and 315. All these multiples split the four quadrants the first time around. Although lots more multiples of 15 exist, the more-important ones end in a 0. Also, the 0-degree angle is among the favorites. A measure of 0 degrees is technically a multiple of any of these measures, and you need it because it's the starting point.

Determining the exact values of functions

Even though a scientific calculator gives you the values of the trig functions of any angle, not just your favorite angles, the values it shows you for most of those angles are just estimates. For example, the exact value of the sine of 60 degrees is $\frac{\sqrt{3}}{2}$. However, because radicals of numbers that aren't perfect squares are irrational and have an endless decimal value, a calculator carries that value out to a certain number of decimal places and then rounds it off. In this case, $\frac{\sqrt{3}}{2} \approx 0.8660254038$. This decimal has many more places than you usually need — normally, three decimal places is enough.

In trig, you frequently use the exact values of the most favorite angles because they give better results in computations and applications, so memorizing those exact values is a good idea.

The process for constructing a table, which I explain in this section, is easy to remember, so you can create one quickly when you need to — either on paper or in your head.

A quick table for the three basic trig functions

The angles used most often in trig have trig functions with convenient exact values. Other angles don't cooperate like these popular ones do.

A quick, easy way to memorize the exact trig-function values of the most common angles is to construct a table, starting with the sine function and working with a pattern of fractions and radicals. Create a table with the top row listing the angles, as shown in Figure 7-6. The first function, in the next row, is sine.

Figure 7-6:
Constructing
a table of
exact
values.

$\theta°$	0°	30°	45°	60°	90°
$\sin\theta$					

The entries following sin θ in the second row are the fractions and radicals with the following pattern:

- ✔ Each fraction has a denominator of 2.
- ✔ The numerators of the fractions are radicals with 0, 1, 2, 3, and 4 under them, in that order, as shown in Figure 7-7.

Figure 7-7:
Creating entries for the second row of the table of exact values.

$\theta°$	0°	30°	45°	60°	90°
sin θ	$\frac{\sqrt{0}}{2}$	$\frac{\sqrt{1}}{2}$	$\frac{\sqrt{2}}{2}$	$\frac{\sqrt{3}}{2}$	$\frac{\sqrt{4}}{2}$

Next, simplify the fractions that can be simplified so the table becomes what you see in Figure 7-8:

$$\frac{\sqrt{0}}{2} = 0, \ \frac{\sqrt{1}}{2} = \frac{1}{2}, \ \frac{\sqrt{2}}{2} = \frac{\sqrt{2}}{2}, \ \frac{\sqrt{3}}{2} = \frac{\sqrt{3}}{2}, \ \frac{\sqrt{4}}{2} = 1$$

Figure 7-8:
The first row of the table with simplified values.

$\theta°$	0°	30°	45°	60°	90°
sin θ	0	$\frac{1}{2}$	$\frac{\sqrt{2}}{2}$	$\frac{\sqrt{3}}{2}$	1

The next row, for the cosine, is just the sine's row in reverse order, as Figure 7-9 shows.

$\theta°$	0°	30°	45°	60°	90°
sin θ	0	$\frac{1}{2}$	$\frac{\sqrt{2}}{2}$	$\frac{\sqrt{3}}{2}$	1
cos θ	1	$\frac{\sqrt{3}}{2}$	$\frac{\sqrt{2}}{2}$	$\frac{1}{2}$	0

Figure 7-9:
Adding the cosine row.

The next row is for the tangent. In a right triangle, you find the tangent of an acute angle with the ratio $\frac{\text{opposite}}{\text{adjacent}}$ (refer to "The tangent function: Opposite over adjacent," earlier in this chapter). You get the same ratio when you divide sine by cosine. Here's how it works:

$$\frac{\text{sine}}{\text{cosine}} = \frac{\dfrac{\text{opposite}}{\text{hypotenuse}}}{\dfrac{\text{adjacent}}{\text{hypotenuse}}} = \frac{\text{opposite}}{\cancel{\text{hypotenuse}}} \cdot \frac{\cancel{\text{hypotenuse}}}{\text{adjacent}} = \frac{\text{opposite}}{\text{adjacent}}$$

Because you already know the values for sine and cosine, you can use this property (tangent equals sine divided by cosine) to get the tangent values for the table:

✔ For the tangent of 0 degrees, $\frac{0}{1} = 0$.

✔ The tangent of 30 degrees is $\dfrac{\frac{1}{2}}{\frac{\sqrt{3}}{2}} = \frac{1}{2} \cdot \frac{2}{\sqrt{3}} = \frac{1}{\sqrt{3}} = \frac{\sqrt{3}}{3}$.

✔ The tangent of 45 degrees is $\dfrac{\frac{\sqrt{2}}{2}}{\frac{\sqrt{2}}{2}} = 1$.

✔ The tangent of 60 degrees is $\dfrac{\frac{\sqrt{3}}{2}}{\frac{1}{2}} = \frac{\sqrt{3}}{2} \cdot \frac{2}{1} = \frac{\sqrt{3}}{1} = \sqrt{3}$.

✔ The tangent of 90 degrees is $\frac{1}{0}$, which is undefined. So the tangent of 90 degrees doesn't have a value — it simply doesn't exist.

See Figure 7-10 for the completed table with the tangent row.

$\theta°$	0°	30°	45°	60°	90°
$\sin\theta$	0	$\frac{1}{2}$	$\frac{\sqrt{2}}{2}$	$\frac{\sqrt{3}}{2}$	1
$\cos\theta$	1	$\frac{\sqrt{3}}{2}$	$\frac{\sqrt{2}}{2}$	$\frac{1}{2}$	0
$\tan\theta$	0	$\frac{\sqrt{3}}{3}$	1	$\sqrt{3}$	undefined

A quick table for the three reciprocal trig functions

If you read "Taking It a Step Further: Reciprocal Functions," earlier in this chapter, you know that the reciprocal functions have values that are reciprocals, or flips, of the values for their respective functions. The reciprocal of sine is cosecant, so each function value is the reciprocal of sine's. The same goes for the other two reciprocal functions. The table in Figure 7-11 shows the reciprocal in each case, in their simplified forms. Whenever you see *undefined,* it's because the original function value was 0, and the reciprocal of 0 has no value.

Figure 7-11:
The reciprocal functions.

$\theta°$	0°	30°	45°	60°	90°
$\csc\theta$	undefined	2	$\sqrt{2}$	$\dfrac{2\sqrt{3}}{3}$	1
$\sec\theta$	1	$\dfrac{2\sqrt{3}}{3}$	$\sqrt{2}$	2	undefined
$\cot\theta$	undefined	$\sqrt{3}$	1	$\dfrac{\sqrt{3}}{3}$	0

The first pocket calculator

Nowadays, many people take for granted the small, efficient, inexpensive pocket calculators that are available. The gadgets do everything imaginable. Before the mid-1940s, though, pocket calculators didn't even exist. Back in 1855, Thomas de Colmar built a machine that could add, subtract, multiply, and divide. The only problem was that it was the size of a piano and weighed more than 60 pounds.

Eighty years later, the Marchant brothers, Rodney and Alfred, along with their chief designer, Carl Friden, tried to shrink the large desk models of the day but still had machines that weighed too much. The Marchant light-weight adding machine weighed 34 pounds, had 9 columns of buttons, and had a carriage with 18 mechanical readouts. It also had two big cranks sprouting from the side. Accountants hauled these "portable" machines around in suitcases.

The Curta, developed by Curt Herzstark, was a calculating machine that hit the market in 1947. It added, subtracted, multiplied, divided, performed square and cube roots, and more. The machine resembled a pepper grinder and was entirely mechanical, with no battery, keyboard, or liquid-crystal display. Like a pepper grinder, you turned the crank on the top, and the results appeared in small windows circling the top. The Curta had the precision of a Swiss watch and was the first truly compact, portable calculator. Herzstark completed the design for this precision calculator while imprisoned in a Buchenwald concentration camp during World War II. During the 1950s and '60s, engineers, scientists, and accountants used the Curta, until the advent of electronic calculators in the '70s.

Chapter 8

Trading Triangles for Circles: Circular Functions

*O*ne of the ways that mathematicians first defined the trig functions was by using ratios formed from the measures of the sides of right triangles (see Chapter 7). Right triangles and the measures of their sides are convenient and easy to construct. This fact led to a sort of natural development of the functions, and it proved to be most useful because it allowed engineers, astronomers, and mathematicians to make accurate calculations of the heights of tall objects, areas of large expanses, and predictions of eclipses and other astronomical phenomena. But, of course, they couldn't stop there. The world of trigonometry and its applications opened up even more when they expanded the trig functions and properties to angles of *any* measure — positive and negative — not just those limited to a right triangle. This extension allowed them to calculate the areas of triangles containing obtuse angles and navigational plots. The best place to begin describing these new functions and comparing them is with the most basic of all circles — the unit circle.

Getting Acquainted with the Unit Circle

The *unit circle* is a circle with its center at the origin of the coordinate plane and with a radius of 1 unit. Any circle with its center at the origin has the equation $x^2 + y^2 = r^2$, where r is the radius of the circle. In the case of a unit circle, the equation is $x^2 + y^2 = 1$. This equation shows that the points lying on the unit circle have to have coordinates (*x*- and *y*-values) that, when you square each of them and then add those values together, equal 1. The coordinates for the points lying on the unit circle and also on the axes are (1,0), (–1,0), (0,1), and (0,–1). These four points (called *intercepts*) are shown in Figure 8-1.

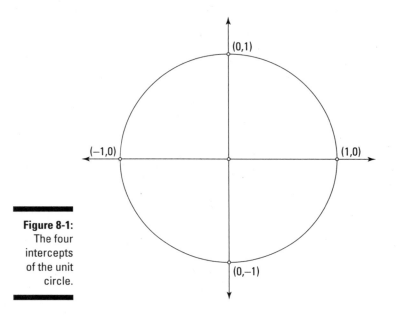

Figure 8-1:
The four
intercepts
of the unit
circle.

Placing points on the unit circle

The rest of the points on the unit circle aren't as nice and neat as those you see in Figure 8-1. They all have fractions or radicals — or both — in them. For instance, the point $\left(\frac{1}{2}, \frac{\sqrt{3}}{2}\right)$ lies on the unit circle. Look at how these coordinates work in the equation of the unit circle:

$$\left(\frac{1}{2}\right)^2 + \left(\frac{\sqrt{3}}{2}\right)^2 = \frac{1}{4} + \frac{3}{4} = 1$$

If you square each coordinate and add those values together, you get 1.

Any combination of these two coordinates, whether the coordinates are positive or negative, gives you a different point on the unit circle. They all work because whether a number is positive or negative, its square is the same positive number. Here are some combinations of those two coordinates that satisfy the unit-circle equation:

$$\left(\frac{1}{2}, \frac{\sqrt{3}}{2}\right), \left(-\frac{1}{2}, \frac{\sqrt{3}}{2}\right), \left(\frac{1}{2}, -\frac{\sqrt{3}}{2}\right), \left(-\frac{\sqrt{3}}{2}, \frac{1}{2}\right), \left(-\frac{\sqrt{3}}{2}, -\frac{1}{2}\right)$$

Another pair of coordinates that works on the unit circle is $\left(\frac{\sqrt{2}}{2}, \frac{\sqrt{2}}{2}\right)$, because the sum of the squares is equal to 1: $\left(\frac{\sqrt{2}}{2}\right)^2 + \left(\frac{\sqrt{2}}{2}\right)^2 = \frac{2}{4} + \frac{2}{4} = \frac{4}{4} = 1$. The numbers that continually crop up as coordinates of points on the unit circle are $0, \frac{1}{2}, \frac{\sqrt{2}}{2}, \frac{\sqrt{3}}{2}$, and 1. If you read Chapter 7, they should look familiar — they're the sine and cosine values of the most common acute-angle measures. Figure 8-2 shows the locations of those points on the unit circle.

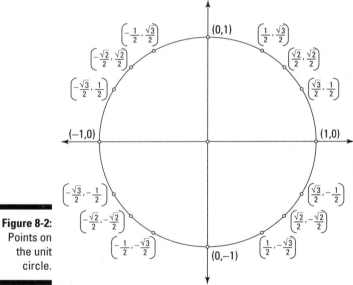

Figure 8-2:
Points on
the unit
circle.

The points on the unit circle shown in Figure 8-2 are frequently used in trigonometry and other math applications, but they aren't the only points on that circle. Every circle has an infinite number of points with all sorts of interesting coordinates — even more interesting than those already shown. If you're looking for the coordinates of some other point on the unit circle, you can just pick some number between –1 and 1 to be the *x*- or the *y*-value and then solve for the other value. I describe this method for finding the other part of a coordinate in the next section. All these other coordinates come into play when you're drawing a ray that starts at the unit circle's center and want to find the trig functions of the angle formed by that ray and the positive *x*-axis.

Babylonian mathematics

The Babylonians were of an ancient culture that had a good deal of influence on the development of mathematics in many areas:

✔ **They developed a system of written symbols.** Way back in the day, the Babylonians developed a form of writing that was based on wedge-shaped symbols called *cuneiform*. Their work has been preserved in the clay tablets that they wrote on, but this way of writing was really cumbersome, so they couldn't write very fast or for very long.

✔ **They gave us the time of day.** The Babylonians divided the day into 24 hours, each hour into 60 minutes, and each minute into 60 seconds. This division resulted in their base-60 counting system — called the *hexasegimal* system — which means that their number system contained 60 different characters, whereas ours has only 10 (thank goodness). It took a long time to learn all these characters, but not everyone was schooled in mathematics or even reading, for that matter.

✔ **They beat Pythagoras to the punch.** These ancients had a knowledge of trigonometry and the Pythagorean theorem 1,000 years before Pythagoras did; they just didn't get the credit he did.

✔ **They were neat freaks.** Their methods for solving problems were very logical and systematic. They preferred orderly procedures based on tables and facts, which is probably why they knew about pi and could approximate its value.

The Babylonians thought in terms of algebra and trigonometry, but you probably wouldn't recognize their notes if you were to pick up a tablet or two to read what they discovered. You'll find this book much more readable!

Finding a missing coordinate

If you have the value of one of a point's coordinates on the unit circle and need to find the other, then you can substitute the known value into the unit-circle equation and solve for the missing value.

You can choose any number between 1 and −1, because that's how far the unit circle extends along the *x*- and *y*-axes. For example, say $\frac{2}{5}$ is the *x*-coordinate of a point on the unit circle. You can find the *y*-coordinate like so:

1. **Substitute the *x*-coordinate value into the unit-circle equation.**

$$\left(\frac{2}{5}\right)^2 + y^2 = 1$$

2. **Square the *x*-coordinate and subtract that value from each side.**

$$\frac{4}{25} + y^2 = 1$$
$$y^2 = 1 - \frac{4}{25} = \frac{21}{25}$$

3. **Take the square root of each side.**

$$\sqrt{y^2} = \pm\sqrt{\frac{21}{25}}$$

$$y = \pm\frac{\sqrt{21}}{5}$$

Note that the y-coordinate can have two values, because the unit circle has two different points for every particular x-coordinate (and for every y-coordinate). Look at Figure 8-3, and you can see how that happens.

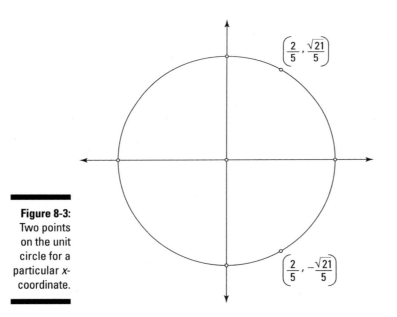

Figure 8-3:
Two points on the unit circle for a particular x-coordinate.

Another example: Find the x-coordinate (or coordinates) if the y-coordinate is $-\frac{7}{25}$.

1. **Substitute the y-coordinate value into the unit-circle equation.**

$$x^2 = \left(-\frac{7}{25}\right)^2 = 1$$

2. **Square the y-coordinate and subtract that value from each side.**

$$x^2 + \frac{49}{625} = 1$$

$$x^2 = 1 - \frac{49}{625} = \frac{576}{625}$$

3. Take the square root of each side.

$$\sqrt{x^2} = \pm\sqrt{\frac{576}{625}}$$

$$x = \pm\frac{24}{25}$$

As you can see, the x-coordinate here has two values, and the two points are $\left(-\frac{7}{25}, \frac{24}{25}\right)$ and $\left(-\frac{7}{25}, -\frac{24}{25}\right)$.

Going Full Circle with the Angles

The unit circle is the platform for describing all the possible angle measures from 0 to 360 degrees, all the negatives of those angles, plus all the multiples of the positive and negative angles from negative infinity to positive infinity. In other words, the unit circle consists of all the angles that exist. Because a right triangle can only measure angles of 90 degrees or less, the circle allows for a much-broader range.

Staying positive

The positive angles on the unit circle are measured with the initial side on the positive x-axis and the terminal side moving counterclockwise around the origin (to figure out which side is which, see Chapter 4). Figure 8-4 shows some positive angles measured in both degrees and radians.

In Figure 8-4, notice that the terminal sides of the angles measuring 30 degrees and 210 degrees, 60 degrees and 240 degrees, and so on form straight lines. This fact is to be expected because the angles are 180 degrees apart, and a straight angle measures 180 degrees. You see the significance of this point when you deal with the trig functions for these angles in "Defining Trig Functions for All Angles," later in this chapter.

Being negative or multiplying your angles

Just when you thought that 360 degrees or 2π radians was enough for anyone, you're confronted with the reality that many of the basic angles have negative values and even multiples of themselves. If you measure angles clockwise instead of counterclockwise, then the angles have negative measures: A 30-degree angle is the same as a −330-degree angle, because they have the same terminal side. Likewise, an angle of $\frac{7\pi}{3}$ is the same as an angle of $-\frac{\pi}{3}$. For rules on how to compute these angles in radian measure, refer to Chapter 1.

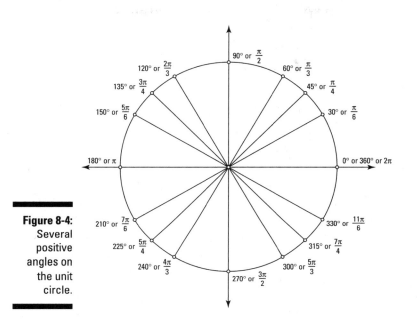

Figure 8-4:
Several positive angles on the unit circle.

But wait — you have even more ways to name an angle. By doing a complete rotation or two (or more) and adding or subtracting 360 degrees or a multiple of it before settling on the angle's terminal side, you can get an infinite number of angle measures, both positive and negative, for the same basic angle. For example, an angle of 60 degrees has the same terminal side as that of a 420-degree angle and a –300-degree angle. Figure 8-5 shows many names for the same 60-degree angle in both degrees and radians.

Although this name-calling of angles may seem pointless at first, there's more to it than arbitrarily using negatives or multiples of angles just to be difficult. The angles that are related to one another have trig functions that are also related, if not the same (more on that in the section "Defining Trig Functions for All Angles").

Locating and computing reference angles

Each of the angles in a unit circle has a *reference* angle, which is always a positive acute angle. By identifying the reference angle, you can determine the function values for that reference angle and, ultimately, the original angle. Sometimes, solving for the reference angle first is much easier than trying to solve for the original angle. The trig functions have values that repeat over and over; sometimes those values are positive, and sometimes they're negative. Using a reference angle helps keep the number of different values to a minimum. You just assign the positive or negative sign after determining which reference angle to match the original angle with.

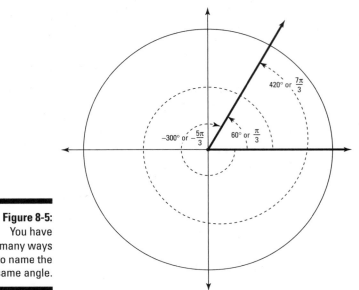

Figure 8-5:
You have
many ways
to name the
same angle.

You can determine a reference angle by looking at the terminal side of the angle you're working with and the positive or negative *x*-axis, depending on which quadrant the terminal side is in:

✔ **Quadrant I (QI):** The reference angle is the same as the original angle itself.

✔ **Quadrant II (QII):** The reference angle is the measure from the terminal side down to the negative *x*-axis.

✔ **Quadrant III (QIII):** The reference angle is the measure from the negative *x*-axis down to the terminal side.

✔ **Quadrant IV (QIV):** The reference angle is the measure from the terminal side up to the positive *x*-axis.

Figure 8-6 shows the positions of the reference angles in the four quadrants.

As with all angles, you can measure reference angles in degrees or radians. I have to admit that I sometimes prefer to work in degrees and will convert a radian measure to do these computations. Whichever method you choose is fine — it's all a matter of taste.

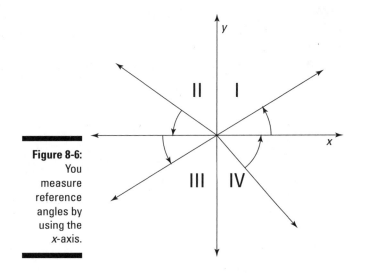

Figuring the angle measure in degrees

To compute the measure (in degrees) of the reference angle for any given angle θ, use the rules in Table 8-1.

Table 8-1	Finding Reference Angles in Degrees	
Quadrant	*Measure of Angle θ*	*Measure of Reference Angle*
I	0° to 90°	θ
II	90° to 180°	$180° - \theta$
III	180° to 270°	$\theta - 180°$
IV	270° to 360°	$360° - \theta$

Using Table 8-1, find the reference angle for 200 degrees:

1. Determine the quadrant in which the terminal side lies.

A 200-degree angle is between 180 and 270 degrees, so the terminal side is in QIII.

2. Do the operation indicated for that quadrant.

Subtract 180 degrees from the angle, which is 200 degrees. You find that $200 - 180 = 20$, so the reference angle is 20 degrees.

Now find the reference angle for 350 degrees:

1. Determine the quadrant in which the terminal side lies.

A 350-degree angle is between 270 and 360 degrees, so the terminal side is in QIV.

2. Do the operation indicated for that quadrant.

Subtract 350 degrees from the angle, which is 360 degrees. You find that 360 − 350 = 10, so the reference angle is 10 degrees.

Sometimes angle measures don't fit neatly in the ranges shown in Table 8-1. For example, you may need to find the reference angle for a negative angle or a multiple of an angle.

To find the reference angle for −340 degrees:

1. Determine the quadrant in which the terminal side lies.

A −340-degree angle is equivalent to a 20-degree angle. (You get the positive angle measure by adding 360, or one full revolution around the origin, to the negative measure.) A 20-degree angle has its terminal side in QI.

2. Do the operation indicated for that quadrant.

Angles in the first quadrant are their own reference angle, so the reference angle is 20 degrees.

On the other end of the spectrum, to find the reference angle for 960 degrees:

1. Determine the quadrant in which the terminal side lies.

A 960-degree angle is equivalent to a 240-degree angle. (You get this measure by subtracting 360 from 960 twice.) A 240-degree angle is between 180 and 270 degrees, so its terminal side is in QIII.

2. Do the operation indicated for that quadrant.

Subtract 180 from 240. You find that 240 − 180 = 60, so the reference angle is 60 degrees.

Figuring the angle measure in radians

To compute the measure (in radians) of the reference angle for any given angle θ, use the rules in Table 8-2.

Table 8-2	Finding Reference Angles in Radians	
Quadrant	*Measure of Angle θ*	*Measure of Reference Angle*
I	0 to $\frac{\pi}{2}$	θ
II	$\frac{\pi}{2}$ to π	$\pi - \theta$
III	π to $\frac{3\pi}{2}$	$\theta - \pi$
IV	$\frac{3\pi}{2}$ to 2π	$2\pi - \theta$

To find the reference angle for $\frac{15\pi}{16}$, for example:

1. **Determine the quadrant in which the terminal side lies.**

   An angle measuring $\frac{15\pi}{16}$ has its terminal side in QII, which you know because $\frac{15}{16}$ is slightly less than 1, making the angle slightly less than π.

2. **Do the operation indicated for that quadrant.**

   Subtract $\frac{15\pi}{16}$ from π. When you do so, you get $\pi - \frac{15\pi}{16} = \frac{16\pi}{16} - \frac{15\pi}{16} = \frac{\pi}{16}$, so the reference angle is $\frac{\pi}{16}$.

Try on another example for size. If you want to find the reference angle for $\frac{40\pi}{3}$:

1. **Determine the quadrant in which the terminal side lies.**

   According to Table 8-2, an angle of $\frac{40\pi}{3}$ has its terminal side in QIII. To determine this answer, subtract 2π (which is one full revolution around the origin, or the equivalent of 360 degrees) from $\frac{40\pi}{3}$ six times. This calculation leaves $\frac{4\pi}{3}$, and that angle has its terminal side in QIII.

2. **Do the operation indicated for that quadrant.**

   Subtract π from $\frac{4\pi}{3}$. In doing so, you get $\frac{4\pi}{3} - \pi = \frac{4\pi}{3} - \frac{3\pi}{3} = \frac{\pi}{3}$, so the reference angle is $\frac{\pi}{3}$.

Defining Trig Functions for All Angles

So many angles are used in trigonometry and other math areas, and the majority of those angles are multiples of 30 and 45 degrees. So having a trick up your sleeve so you can quickly access the function values of this frequent-flier list of angles makes perfect sense. All you need to know are the values of the trig functions for 0-, 30-, 45-, 60-, and 90-degree angles in order to determine *all* the trig functions of *all* the angles, positive or negative, that are multiples of 30 or 45 degrees, which are the two most basic, foundational angles. Finding these function values is a two-step process. First, find the measure of the reference angle, and second, determine whether the function value is positive or negative.

Putting reference angles to use

The first step to finding the function value of one of the angles that's a multiple of 30 or 45 degrees is to find the reference angle. When the reference angle comes out to be 0, 30, 45, 60, or 90 degrees, you can use the function value of that angle and then figure out the sign (see the next section). Use Table 8-1 or Table 8-2, earlier in this chapter, to find the reference angle.

All angles with a 30-degree reference angle have trig functions whose absolute values are the same as those of the 30-degree angle. The sines of 30, 150, 210, and 330 degrees, for example, are all either $\frac{1}{2}$ or $-\frac{1}{2}$. Likewise, using a 45-degree angle as a reference angle, the cosines of 45, 135, 225, and 315 degrees, for example, are all $\frac{\sqrt{2}}{2}$ or $-\frac{\sqrt{2}}{2}$.

Labeling the optimists and pessimists

The sine values for 30, 150, 210, and 330 degrees are, respectively, $\frac{1}{2}$, $\frac{1}{2}$, $-\frac{1}{2}$, and $-\frac{1}{2}$. All these multiples of 30 degrees have an absolute value of ½ (as I explain in the preceding section). The following rule and Figure 8-7 help you determine whether a trig-function value is positive or negative. First, note that each quadrant in the figure is labeled with a letter. The letters aren't random; they stand for trig functions.

Reading around the quadrants, starting with QI and going counterclockwise, the rule goes like this: If the terminal side of the angle is in the quadrant with letter

> ✓ **A:** All functions are positive
>
> ✓ **S:** Sine and its reciprocal, cosecant, are positive
>
> ✓ **T:** Tangent and its reciprocal, cotangent, are positive
>
> ✓ **C:** Cosine and its reciprocal, secant, are positive

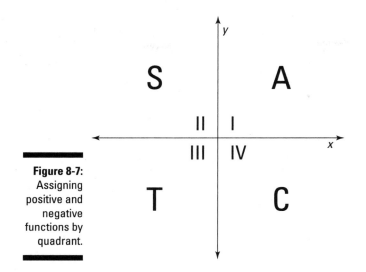

Figure 8-7:
Assigning
positive and
negative
functions by
quadrant.

In QII, only sine and cosecant are positive. All the other function values for angles in this quadrant are negative — and the rule continues in like fashion for the other quadrants.

My trig teacher, Dr. Johnson, showed me a great way to remember this rule: "All Students Take Calculus." (And so I did!) If math is already giving you nightmares, maybe you'd prefer "Any Snake Teases Chickens" or "Apple Sauce Turns Colors." Make up your own! Have at it!

Combining all the rules

Using the rules for reference angles, the values of the functions of certain acute angles (see Chapter 7), and the rule for the signs of the functions, you can determine the trig functions for any angles that relate to the unit circle — any that are graphed in *standard position* (meaning the vertex of the angle is at the origin, and the initial side lies along the positive x-axis). Figure 8-8 combines information from this chapter and the preceding chapter to give you the information you need.

Now, armed with all the necessary information, find the tangent of 300 degrees.

1. Find the reference angle.

Using Figure 8-8, you can see that a 300-degree angle is in the fourth quadrant, so you find the reference angle by subtracting 300 from 360. Therefore, the measure of the reference angle is 60 degrees.

Quadrant	Measure of Angle θ	Measure of Reference Angle
I	0° to 90°	θ
II	90° to 180°	$180° - \theta$
III	180° to 270°	$\theta - 180°$
IV	270° to 360°	$360° - \theta$

$\theta°$	0°	30°	45°	60°	90°
$\sin \theta$	0	$\dfrac{1}{2}$	$\dfrac{\sqrt{2}}{2}$	$\dfrac{\sqrt{3}}{2}$	1
$\cos \theta$	1	$\dfrac{\sqrt{3}}{2}$	$\dfrac{\sqrt{2}}{2}$	$\dfrac{1}{2}$	0
$\tan \theta$	0	$\dfrac{\sqrt{3}}{3}$	1	$\sqrt{3}$	undefined

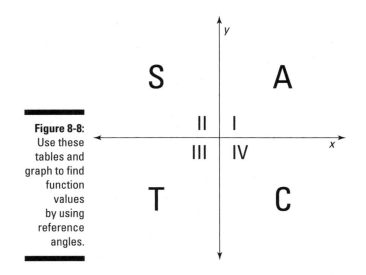

Figure 8-8:
Use these tables and graph to find function values by using reference angles.

2. **Find the numerical value of the tangent.**

   Using Figure 8-8, you see that the numerical part of the tangent of 60 degrees is $\sqrt{3}$.

3. **Find the sign of the tangent.**

   Because a 300-degree angle is in the fourth quadrant, and angles in that quadrant have negative tangents (refer to the section "Labeling the optimists and pessimists"), the tangent of 300 degrees is $-\sqrt{3}$.

To try your hand at working with radians, find the cosecant of $\frac{7\pi}{6}$.

1. **Find the reference angle.**

   Using Figure 8-8, this angle is in the third quadrant, so you find the reference angle by subtracting π from $\frac{7\pi}{6}$, resulting in $\frac{7\pi}{6} - \pi = \frac{\pi}{6}$.

2. **Find the numerical value of the cosecant.**

   In Figure 8-8, the cosecant doesn't appear. However, the reciprocal of the cosecant is sine. So find the value of the sine, and use its reciprocal. The sine of $\frac{\pi}{6}$ is $\frac{1}{2}$, which means that the cosecant is 2.

3. **Find the sign of the cosecant.**

   In the third quadrant, the cosecant of an angle is negative (refer to the section "Labeling the optimists and pessimists"), so the cosecant of $\frac{7\pi}{6}$ is -2.

Using Coordinates of Circles to Solve for Trig Functions

Another way to find the values of the trig functions for angles is to use the coordinates of points on a circle that has its center at the origin. Letting the positive x-axis be the initial side of an angle, you can use the coordinates of the point where the terminal side intersects with the circle to determine the trig functions. Figure 8-9 shows a circle with a radius of r that has an angle drawn in standard position.

The equation of the circle is $x^2 + y^2 = r^2$ (flip back to Chapter 2 for a refresher). Based on this equation and the coordinates of the point where the terminal side of the angle intersects the circle, the six trig functions for angle θ are defined as follows:

$$\sin \theta = \frac{y}{r} \qquad\qquad \csc \theta = \frac{r}{y}$$

$$\cos \theta = \frac{x}{r} \qquad\qquad \sec \theta = \frac{r}{x}$$

$$\tan \theta = \frac{y}{x} \qquad\qquad \cot \theta = \frac{x}{y}$$

You can see where these definitions come from if you picture a right triangle formed by dropping a perpendicular segment from the point (x,y) to the x-axis. Figure 8-10 shows such a right triangle. Remember that the x-value is to the right (or left) of the origin, and the y-value is above (or below) the x-axis — and use those values as lengths of the triangle's sides. Therefore, the side opposite angle θ is y, the value of the y-coordinate. The adjacent side is x, the value of the x-coordinate.

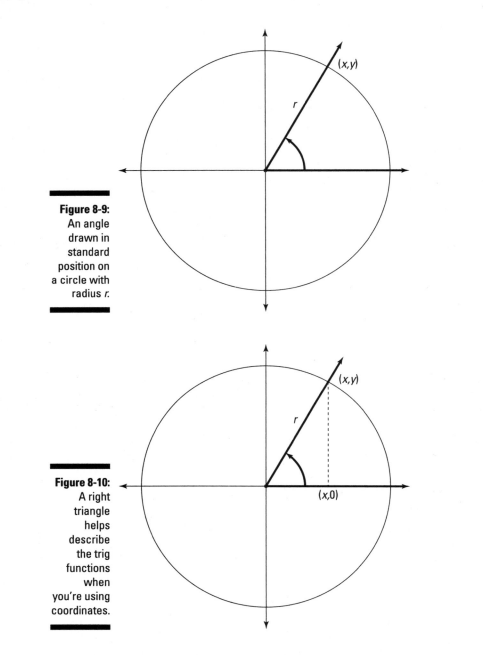

Figure 8-9:
An angle drawn in standard position on a circle with radius *r.*

Figure 8-10:
A right triangle helps describe the trig functions when you're using coordinates.

Take note that for angles in the second quadrant, for example, the *x*-values are negative, and the *y*-values are positive. The radius, however, is always a positive number. With the *x*-values negative and the *y*-values positive, using the definitions for the functions listed earlier in this section, you see that the sine and cosecant are positive, but the other functions are all negative, because they all have an *x* in their ratios. The signs of the trig functions all fall into line when you use this coordinate system, so no need to worry about remembering the ASTC rule here. (For more on that rule and when to use it, see "Labeling the optimists and pessimists.")

Calculating with coordinates on the unit circle

Calculating trig functions on angles within a unit circle is easy as pie. Figure 8-11 shows a unit circle, which has the equation $x^2 + y^2 = 1$, along with some points on the circle and their coordinates.

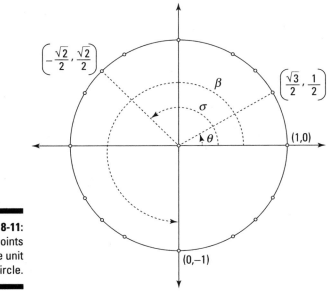

Figure 8-11: Some points on the unit circle.

Using the angles in Figure 8-11, find the tangent of θ.

1. **Find the x- and y-coordinates of the point where the angle's terminal side intersects with the circle.**

   The coordinates are $x = \frac{\sqrt{3}}{2}$ and $y = \frac{1}{2}$. The radius is $r = 1$.

2. **Determine the ratio for the function and substitute in the values.**

   Tangent is $\frac{y}{x}$, so you find that

   $$\frac{y}{x} = \frac{\frac{1}{2}}{\frac{\sqrt{3}}{2}} = \frac{1}{2} \cdot \frac{2}{\sqrt{3}} = \frac{1}{\sqrt{3}} \cdot \frac{\sqrt{3}}{\sqrt{3}} = \frac{\sqrt{3}}{3}$$

Next, using the angles in Figure 8-11, find the cosine of σ.

1. **Find the x- and y-coordinates of the point where the terminal side of the angle intersects with the circle.**

   The coordinates are $x = -\frac{\sqrt{2}}{2}$ and $y = \frac{\sqrt{2}}{2}$; the radius is $r = 1$.

2. **Determine the ratio for the function and substitute in the values.**

   Cosine is $\frac{x}{r}$, which means that you need only the x-coordinate, so

   $$\frac{x}{r} = \frac{-\frac{\sqrt{2}}{2}}{1} = -\frac{\sqrt{2}}{2}.$$

Now, using the angles in Figure 8-11, find the cosecant of β.

1. **Find the x- and y-coordinates of the point where the terminal side of the angle intersects with the circle.**

   The coordinates are $x = 0$ and $y = -1$; the radius is $r = 1$.

2. **Determine the ratio for the function and substitute in the values.**

   Cosecant is $\frac{r}{y}$, which means that you need only the y-coordinate, so
   $\frac{r}{y} = \frac{1}{-1} = -1.$

Calculating with coordinates on any circle at the origin

You don't have to have a unit circle to use this coordinate business when determining the function values of angles graphed in standard position on a circle. You can use a circle with any radius, as long as the center is at the origin.

Using the angles in Figure 8-12, find the sine of α.

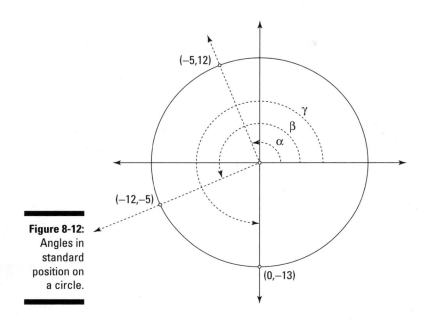

Figure 8-12:
Angles in
standard
position on
a circle.

1. **Find the *x*- and *y*-coordinates of the point where the terminal side of the angle intersects with the circle.**

   The coordinates are $x = -5$ and $y = 12$.

2. **Determine the radius of the circle.**

   The equation of a circle with its center at the origin is $x^2 + y^2 = r^2$. Replacing the *x* and *y* in this equation with –5 and 12, respectively, you get $(-5)^2 + (12)^2 = 25 + 144 = 169 = r^2$. The square root of 169 is 13, so the radius is 13.

3. **Determine the ratio for the function and substitute in the values.**

   The ratio for sine is $\frac{y}{r}$, which means that you need only the *y*-coordinate and radius, so $\frac{y}{r} = \frac{12}{13}$.

Next, using the angles in Figure 8-12, find the cotangent of β.

1. **Find the *x*- and *y*-coordinates of the point where the terminal side of the angle intersects with the circle.**

   The coordinates are $x = -12$ and $y = -5$.

   The cotangent function uses only the *x*- and *y*-coordinates, so you don't need to solve for the radius.

2. **Determine the ratio for the function and substitute in the values.**

   Cotangent is $\frac{x}{y}$, so $\frac{x}{y} = \frac{-12}{-5} = \frac{12}{5}$.

Ancient math contest

Even in the late 1500s, mathematicians around the world found themselves competing with one another. The Belgian mathematician Adriaen Van Roomen challenged other mathematicians to solve a polynomial equation of the 45th degree. It looked something like this (with variables and numbers in place of the ellipsis, of course): $x^{45} - 45x^{43} + 945x^{41} - \ldots + 45x = k$, where k is some constant. At that time, the mathematicians in Belgium and France found themselves in quite a competition, because Van Roomen suggested that no one in France would be able to solve the equation. But French mathematician François Viète put him to shame: He used trigonometry to solve the puzzle. (He let k be equal to the sine of 45 degrees and applied trig identities to find the positive solutions.)

Now, using the angles in Figure 8-12, find the secant of γ.

1. **Find the x- and y-coordinates of the point where the terminal side of the angle intersects with the circle.**

   The coordinates are $x = 0$ and $y = -13$.

2. **Determine the radius of the circle.**

   Per the first example in this section, the radius is 13.

3. **Determine the ratio for the function and substitute in the values.**

   Secant is $\frac{r}{x}$, so you need only the x-coordinate; substituting in, you get $\frac{r}{x} = \frac{13}{0}$. This answer is undefined, which means that angle γ has no secant.

Defining Domains and Ranges of Trig Functions

The *domain* of a function consists of all the input values that a function can handle — the way the function is defined. Of course, you want to get output values (which make up the *range*) when you enter input values (for the basics on domain and range, see Chapter 3). But sometimes, when you input a fraction or a radical into the function, you end up with some impossible situations. In these cases, you need to limit what you put into the function — the domain has to be restricted. For example, the cosecant is defined as the hypotenuse divided by the opposite side (see Chapter 7). If the terminal side of the angle is on the x-axis, then the opposite side is 0, and you're asked to divide by 0. Impossible!

Trig functions have inputs of angle measures, either in degrees or radians. The outputs of the trig functions are real numbers. The hitch here is that the different trig functions have different domains and ranges. Sine and cosine are very cooperative and have the same domain and range. The tangent function and the reciprocal functions, however, all differ. The best way to describe these different domains and ranges is visually: Refer to the coordinate plane with a circle centered on the origin and a right triangle inside it, formed by dropping a line from any point (x,y) on the circle to the x-axis (see Figure 8-13). Remember that r stands for the radius of the circle (and also the hypotenuse of the right triangle in this figure).When that hypotenuse lies along one of the axes, one of the sides is equal to 0, which is a no-no in the denominator of a fraction.

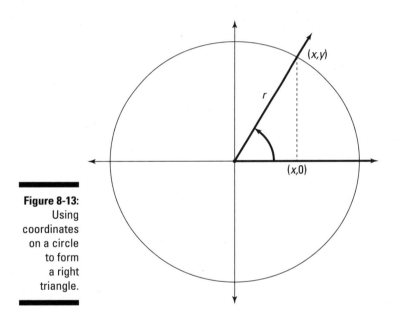

Figure 8-13:
Using coordinates on a circle to form a right triangle.

Consider the values of the variables in Figure 8-13 in relation to one another. The radius, r, is always positive. And the absolute values of x and y (the lengths of the segments they represent) are always smaller than r, unless the point (x, y) is on one of the axes — then one of the values is equal to r and the other is equal to 0.

Friendly functions: Sine and cosine

The sine and cosine functions are unique in the world of trig functions, because their ratios always have a value. The value may be 0, but that's a number, too. In reference to the coordinate plane, sine is $\frac{y}{r}$, and cosine is $\frac{x}{r}$.

The radius, *r*, is always some positive number (which is why these functions always have a value, because they don't ask you to divide by 0), and it's always a number greater than the absolute value of *x* or *y*.

Domains of sine and cosine

The domains of sine and cosine are infinite. In trig speak, you say something like this: If θ represents all the angles in the domain of the two functions $f(\theta)$ = sin θ and $g(\theta)$ = cos θ, then $-\infty \le \theta \le \infty$, which means that θ can be any angle in degrees or radians — any real number.

Ranges of sine and cosine

The output values for sine and cosine are always between –1 and 1. In trig speak, it goes something like this: If $f(\theta)$ and $g(\theta)$ represent the output values of the functions $f(\theta)$ = sin θ and $g(\theta)$ = cos θ, then $-1 \le f(\theta) \le 1$ and $-1 \le g(\theta) \le 1$.

The ratios $\frac{y}{r}$ and $\frac{x}{r}$ will never be improper fractions — the numerator can never be greater than the denominator — because the value of *r*, the radius, is always the biggest number. At best, if the angle θ has a terminal side on an axis (meaning that one of the sides is equal to *r*), then the value of those ratios is 1 or –1.

Close cousins of their reciprocals: Cosecant and secant

The cosecant and secant functions are closely tied to sine and cosine, because they're the respective reciprocals. In reference to the coordinate plane, co-secant is $\frac{r}{y}$, and secant is $\frac{r}{x}$. The value of *r* is the length of the hypotenuse of a right triangle — which, as you find out at the beginning of this section, is always positive and always greater than *x* and *y*. The only problem that arises when computing these functions is when either *x* or *y* is 0 — when the termi-nal side of the angle is on an axis. A function with a 0 in the denominator creates a number or value that doesn't exist (in math speak, the result is *undefined*), so anytime *x* or *y* is 0, you can't perform the cosecant or secant functions. The *x* is 0 when the terminal side is on the *y*-axis, and the *y* is 0 when the terminal side is on the *x*-axis.

Domains of cosecant and secant

The domains of cosecant and secant are restricted — you can only use the functions for angle measures with output numbers that exist.

Anytime the terminal side of an angle lies along the *x*-axis (where *y* = 0), you can't perform the cosecant function on that angle. In trig speak, the rule looks like this: If $h(\theta) = \csc \theta$, then $\theta \neq 0, 180, 360, 540 \ldots$, or any multiple of 180 degrees. In radians, $\theta \neq 0, \pi, 2\pi, 3\pi \ldots$, or any multiple of π.

Anytime the terminal side of an angle lies along the *y*-axis (where *x* = 0), you can't perform the secant function on that angle. So in trig speak, you'd say this: If $k(\theta) = \sec \theta$, then $\theta \neq 90, 270, 450, 630 \ldots$, or any odd multiple of 90 degrees. In radians, $\theta \neq \frac{\pi}{2}, \frac{3\pi}{2}, \frac{5\pi}{2}, \frac{7\pi}{2} \ldots$, or any odd multiple of $\frac{\pi}{2}$.

Ranges of cosecant and secant

The ratios of the cosecant and secant functions on the coordinate plane, $\frac{r}{y}$ and $\frac{r}{x}$, have the hypotenuse, *r*, in the numerator. Because *r* is always positive and greater than or equal to *x* and *y*, these fractions are always *improper* (greater than 1) or equal to 1. The ranges of these two functions never include *proper* fractions (numbers between –1 and 1).

If $h(\theta)$ and $k(\theta)$ are the output values of the functions $h(\theta) = \csc \theta$ and $k(\theta) = \sec \theta$, then $h(\theta) \leq -1$ or $h(\theta) \geq 1$ and $k(\theta) \leq -1$ or $k(\theta) \geq 1$.

Brothers out on their own: Tangent and cotangent

The tangent and cotangent are related not only by the fact that they're reciprocals, but also by the behavior of their ranges. In reference to the coordinate plane, tangent is $\frac{y}{x}$, and cotangent is $\frac{x}{y}$. The domains of both functions are restricted, because sometimes their ratios could have 0s in the denominator, but their ranges are infinite.

Domains of tangent and cotangent

Because *x* can't equal 0 for the tangent function to work, this rule holds true: If $m(\theta) = \tan \theta$, then $\theta \neq 90, 270, 450, 630 \ldots$, or any odd multiple of 90 degrees. In radians, $\theta \neq \frac{\pi}{2}, \frac{3\pi}{2}, \frac{5\pi}{2}, \frac{7\pi}{2} \ldots$, or any odd multiple of $\frac{\pi}{2}$. Both the tangent and secant functions have ratios with *x* in the denominator, making their domains the same.

In order for the cotangent function to work, *y* can't equal 0. If $n(\theta) = \cot \theta$, then $\theta \neq 0, 180, 360, 540 \ldots$, or any multiple of 180 degrees. In radians, $\theta \neq 0, \pi, 2\pi, 3\pi \ldots$, or any multiple of π.

Ranges of tangent and cotangent

The ranges of both tangent and cotangent are infinite, which, when expressed in mathematical notation, looks like this: $-\infty \le m(\theta) \le \infty$ and $-\infty \le n(\theta) \le \infty$.

The range values for these functions get very small (toward negative infinity) or very large (toward positive infinity) whenever the denominator of the respective ratio gets close to 0. When you divide some number by a very small value, such as 0.0001, the result is large. The smaller the denominator, the larger the result.

Chapter 9

Applying Yourself to Trig Functions

*B*ack when trig functions were first developed or recognized — *way* back when — the motivation for creating them wasn't so men could sit around and say, "Hey, Gallius Caesar, did you know that the sine of 45 degrees is $\frac{\sqrt{2}}{2}$?"

Rather, the math gurus of the past worked out the principles of trigonometry because they needed some order or consistency to the numbers that they were applying to astronomy, agriculture, and architecture. They figured out the relationships among all these numbers and shared them with the rest of the known, civilized world.

First Things First: Elevating and Depressing

Mathematical problems that require the use of trig functions often have two common angles: the angle of elevation and the angle of depression. The scenarios that use these angles usually involve, for example, finding the distance from an airplane to a point on the ground or the distance up to a balloon or another object above you. Use the trig functions to solve for the missing part of the ratio or the side of the imaginary right triangle.

Ptolemy: Part right and part wrong

Ptolemy, also known as Claudius Ptolemaeus, was a Greek citizen who lived in Alexandria, Egypt, from about A.D. 87–150. He was an astronomer, mathematician, and geographer. Ptolemy believed that the Sun and other planets revolved around Earth. At that time, only five planets were known, and he believed that they revolved around Earth in this order: Mercury, Venus, the Sun, Mars, Jupiter, and Saturn. This theory was known as the Ptolemaic system. It predicted the positions of the planets with reasonable accuracy, considering that they were naked-eye observations, not enhanced with telescopes. Ptolemy may not have really believed in this system of the planets, though — perhaps he used it only as a method of calculating the relative positions of them. This man did get it right, though, in determining that Earth is a sphere, not flat — this theory affected much of his important work in geography and cartography. His works, and an error that had Asia extending too far to the east, probably influenced Columbus's decision to sail west for the Indies.

Angle of elevation

An *angle of elevation* is measured from the horizontal going upward. The horizontal is usually the ground, street, floor, or any other flat object. Even though the ground isn't perfectly flat or horizontal, you determine the measurements with the assumption that it is. In trig, you have to consider the optimal situation — focus on the big picture, rather than on the imperfections. Figure 9-1 shows an angle of elevation.

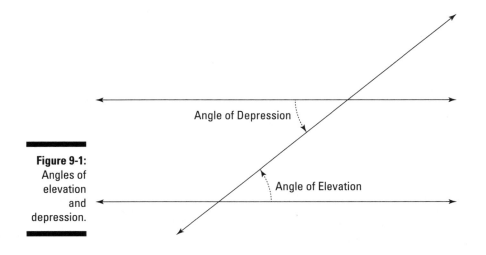

Figure 9-1:
Angles of elevation and depression.

Angle of Depression

Angle of Elevation

Angle of depression

An *angle of depression* is measured from the horizontal going downward. In this case, the horizontal is, for example, an airplane's flight path or a person's line of sight while standing on a mountaintop. The angle is formed when a person on that plane or mountaintop looks at an object on the ground (or on a path parallel to the ground). Refer to Figure 9-1 for a sample angle of depression.

Measuring Tall Buildings with a Single Bound

Every day, people use trigonometry to measure things that they can't reach. How high is that building? Will this ladder reach to the top of that tree? By using the appropriate trig functions, you can find answers to such questions. Two major considerations to keep in mind when working out problems by using trig are as follows: Which trig function should you use, and what are the units or measures in the answer?

The missing values in the ratios of the trig functions represent the missing parts in the problems. You assign the known values appropriately and solve for what's left.

Rescuing a damsel from a tower

Consider the oh-so-common scenario: A damsel is in distress and is being held captive in a tower. Her knight in shining armor is on the ground below with a ladder. He needs to know whether it'll reach her or whether he needs a longer ladder.

When the stunning knight stands 15 feet from the base of the tower and looks up at his precious damsel, the angle of elevation to her window is 60 degrees. How long does the ladder have to be? Figure 9-2 shows the situation in pictorial form.

1. **Identify the parts of the right triangle that you can use to solve the problem.**

   You know that the acute angle is 60 degrees, and the adjacent side of the triangle is along the ground; the distance from the vertex of the angle (where the knight is standing) to the base of the tower is 15 feet. The hypotenuse is the length needed for the ladder — call it *x*. Figure 9-3 shows you the triangle.

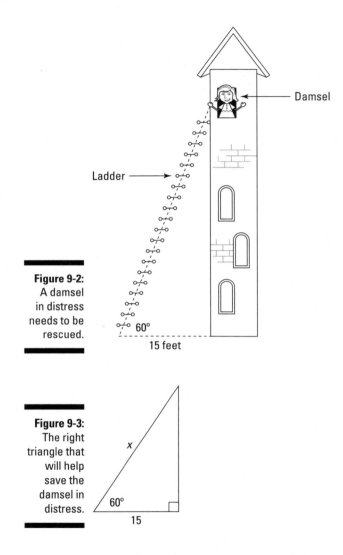

Figure 9-2:
A damsel
in distress
needs to be
rescued.

Figure 9-3:
The right
triangle that
will help
save the
damsel in
distress.

2. Determine which trig function to use.

The adjacent side and hypotenuse are parts of the cosine ratio. Those sides are also parts of the secant ratio, but if at all possible, you should use the three main functions, not their reciprocals.

3. Write an equation with the trig function, then input the values that you know.

For a refresher on those values, look at the charts in Chapter 7.

$$\cos 60° = \frac{\text{adj}}{\text{hyp}}$$

$$\frac{1}{2} = \frac{15}{x}$$

4. Solve the equation.

Cross-multiplying, you get

$$\frac{1}{2} = \frac{15}{x}$$

$$1 \cdot x = 15 \cdot 2$$

$$x = 30$$

The ladder needs to be 30 feet long. (That knight had better be pretty strong!)

Determining the height of a tree

Suppose you're flying a kite, and it gets caught at the top of a tree. You've let out all 100 feet of string for the kite, and the angle that the string makes with the ground (the angle of elevation) is 75 degrees. Instead of worrying about how to get your kite back, you wonder, "How tall is that tree?"

Figure 9-4 shows the scenario.

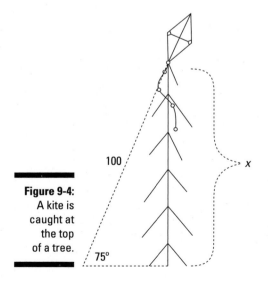

Figure 9-4:
A kite is caught at the top of a tree.

100

x

75°

To find a solution to your quandary, follow these steps:

1. **Identify the parts of the right triangle that you can use to solve the problem.**

   The hypotenuse of the right triangle is the length of the string. The side opposite the 75-degree angle is what you're solving for; call it x.

2. **Determine which trig function to use.**

   The hypotenuse and opposite side are part of the sine ratio.

3. **Write an equation with the trig function, then input the values that you know.**

   The 75-degree angle isn't one of the more-common angles, so use a scientific calculator or one of the tables in the Appendix to obtain a value for the sine, correct to three decimal places.

   $$\sin 75° = \frac{\text{opp}}{\text{hyp}}$$
   $$0.966 = \frac{x}{100}$$

4. **Solve the equation.**

   Cross-multiplying, you get

   $$0.966 = \frac{x}{100}$$
   $$0.966 \cdot 100 = x$$
   $$96.6 = x$$

   The tree is over 96 feet tall.

Measuring the distance between buildings

Jumping Jehoshaphat makes his living by jumping, on his motorcycle, from building to building, cliff to bluff, or any place he can get attention for doing it. His record jump is a distance of 260 feet, from one building to another. Jehoshaphat is on to his next feat and needs to determine the distance from one building to another. His assistant, Lovely Lindsay, holds a 6-foot pole perpendicular to the roof she's standing on. When Jehoshaphat, standing on top of the first building, sights straight across to a point at the base of the pole and then sights a point halfway up the pole, the angle of elevation is 1 degree. Will he be able to make the jump? See Figure 9-5 for a visual of Jehoshaphat's calculation.

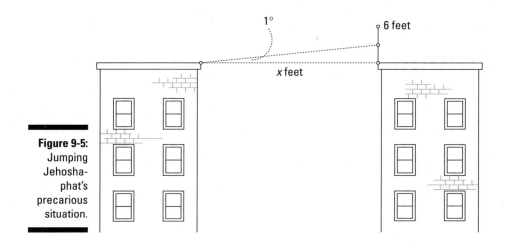

Figure 9-5:
Jumping
Jehosha-
phat's
precarious
situation.

1. **Identify the parts of the right triangle that you can use to solve the problem.**

   You know the value for the side opposite the 1-degree angle, which is half the pole length, and the adjacent side is the unknown distance. Call that distance x.

2. **Determine which trig function to use.**

   The tangent of an angle uses opposite divided by adjacent.

3. **Write an equation with the trig function, then input the values that you know.**

   The length of half the pole is 3 feet, so the equation looks like this:

   $$\tan 1° = \frac{\text{opp}}{\text{adj}} = \frac{3}{x}$$

4. **Solve for the value of x.**

   Use the Appendix to find the value of the tangent of 1 degree.

   $$\tan 1° = \frac{3}{x}$$
   $$0.0175 = \frac{3}{x}$$
   $$x = \frac{3}{0.0175} = 171.4$$

   You find that the distance between the buildings is a little less than 172 feet across. Jehoshaphat should be able to make the jump easily, because his record is 260 feet.

Measuring Slope

Have you ever noticed a worker along the road, peering through an instrument, looking at a fellow worker holding up a sign or flag? Haven't you ever wondered what they're doing? Have you wanted to get out and look through the instrument, too? With trigonometry, you can do just what those workers do — measure distances and angles. Land surveyors use trigonometry and their fancy equipment to measure things like the slope of a piece of land.

If you read the first section in this chapter, you may recognize that the slope of land downward is sort of like an angle of depression. Slopes, angles of depression, and angles of elevation are all interrelated because they use the same trig functions. It's just that in slope applications, you're solving for the angle rather than being given it.

To solve problems involving slope, you can use the trig ratios and right triangles. One side of the triangle is the distance from one worker to the other; the other side is the vertical distance from the ground to a point on a pole. You form a ratio with those measures and determine the angle — voilà!

Suppose that Dan and Fran are making measurements for the road-paving crew. They need to know how much the land slopes downward along a particular stretch of road. Dan walks 80 feet from Fran and holds up a long pole, perpendicular to the ground, that has markings every inch along it. Fran looks at the pole through a sighting instrument. Looking straight across, parallel to the horizon, Fran sights a point on the pole 50 inches above the ground — call it point A. Then Fran looks through the instrument at the bottom of the pole, creating an angle of depression. See Figure 9-6 for a diagram of this situation. What is the angle of depression, or slope of the road, to where Dan is standing?

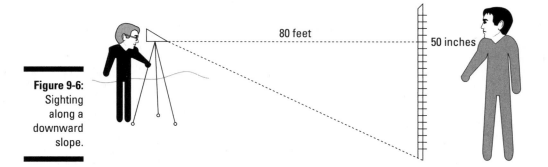

Figure 9-6:
Sighting along a downward slope.

80 feet

50 inches

1. **Identify the parts of the right triangle that you can use to solve the problem.**

   The values you know are for the sides adjacent to and opposite the angle of depression. Call the angle measure x.

2. **Determine which trig function to use.**

   The tangent of the angle with measure x uses opposite divided by adjacent.

3. **Write an equation with the trig function, then input the values that you know.**

   $$\tan x = \frac{\text{opposite}}{\text{adjacent}}$$

   In this problem, you need to write the equation with a common unit of measurement — either feet or inches. Changing 80 feet to inches makes for a big number; changing 50 inches to feet involves a fraction or decimal. Whichever unit you choose is up to you. In this example, I choose the big number, so I convert feet to inches.

   $80 \text{ feet} = 80 \cdot 12 \text{ inches} = 960 \text{ inches}$

   Substituting in the values, you get the tangent of some angle with a measure of x degrees:

   $$\tan x = \frac{\text{opp}}{\text{adj}} = \frac{50}{960} = 0.05208333$$

4. **Solve for the value of x.**

   In the Appendix, you see that an angle of 2.9 degrees has a tangent of 0.0507, and a 3-degree angle has a tangent of 0.0524. The 3-degree angle has a tangent that's closer to 0.05208333, so you can estimate that the road slopes at a 3-degree angle between Fran and Stan.

The Sky's (Not) the Limit

Early trigonometry had many earthbound applications — surveyors and engineers have used it for centuries. Over time, astronomers and navigators on journeys around the world began using trig to solve many mysteries here on earth and in outer space. They estimated or measured angles by sighting objects in the heavens and charting their movements. Then they used the angle between one sighting and another to solve for the distances that are unreachable.

Spotting a balloon

Cindy and Mindy, standing a mile apart, spot a hot-air balloon directly above a particular point on the ground. The angle of elevation from Cindy to the balloon is 60 degrees; the angle of elevation from Mindy to the balloon is 70 degrees. Figure 9-7 shows a visual representation. How high is the balloon?

Figure 9-7:
Two friends
spot a hot-
air balloon.

60°   70°

1 mile

If you look at Figure 9-7, you see that two right triangles are formed. The two triangles share a side — the one opposite the measured acute angle in each. Call its length y. The two adjacent sides add up to 1 mile, so you can keep the variables to a minimum by naming one side x and the other $1-x$. Figure 9-8 shows the triangles with the variables.

To figure out how high the balloon is, follow these steps:

1. **Identify the parts of the triangles that you can use to solve the problem.**

   In both triangles, you have variables for the adjacent and opposite sides of the acute angles of elevation.

2. **Determine which trig function to use.**

   The tangent uses the opposite and adjacent sides.

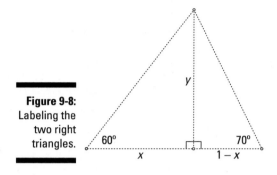

Figure 9-8:
Labeling the
two right
triangles.

3. **Write equations with the trig functions.**

$$\tan 60° = \frac{\text{opp}}{\text{adj}} = \frac{y}{x} \qquad \tan 70° = \frac{\text{opp}}{\text{adj}} = \frac{y}{1-x}$$

and

$$\tan 60° = \frac{y}{x} \qquad\qquad\qquad \tan 70° = \frac{y}{1-x}$$

4. **Solve for x by setting the equations equal to one another.**

Solve each of the equations for y.

$$\tan 60° = \frac{y}{x} \qquad\text{and}\qquad \tan 70° = \frac{y}{1-x}$$

$$x \cdot \tan 60° = y \qquad\qquad (1-x) \cdot \tan 70° = y$$

Set those two equations equal to one another and solve for x.

$$x \cdot \tan 60° = (1-x)\tan 70°$$
$$x \cdot \tan 60° = \tan 70° - x \cdot \tan 70°$$
$$x \cdot \tan 60° + x \cdot \tan 70° = \tan 70°$$
$$x(\tan 60° + \tan 70°) = \tan 70°$$
$$x = \frac{\tan 70°}{\tan 60° + \tan 70°}$$

5. **Solve for the value of x.**

You find the value of x by finding the values of the functions with a calculator or in the Appendix. Upon doing so, you find that x is approximately 0.613 miles. Put that value into one of the equations to solve for y:

$$x \cdot \tan 60° = y$$
$$(0.613)\tan 60° = y$$
$$(0.613)(1.732) = y$$
$$1.062 = y$$

The balloon is 1.062 miles high — sounds a tad high to me!

Tracking a rocket

In this example, a rocket is shot off and travels vertically as a scientist, who's a mile away, watches its flight. One second into the flight, the angle of elevation of the rocket is 30 degrees. Two seconds later, the angle of elevation is 60 degrees. How far did the rocket travel in those two seconds? Figure 9-9 shows the rocket rising vertically.

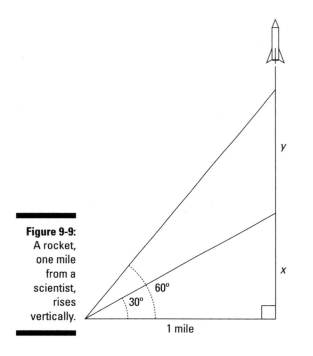

Figure 9-9:
A rocket, one mile from a scientist, rises vertically.

60°

30°

y

x

1 mile

1. **Identify the parts of the triangles that you can use to solve the problem.**

   In Figure 9-9, you can see two right triangles. One is superimposed on the other and shares a side — the adjacent side. In both triangles, the relevant sides are those that are adjacent and opposite the angles of elevation.

2. **Determine which trig function to use.**

   The tangent uses the adjacent and opposite sides.

3. **Write equations with the trig functions.**

$$\tan 30° = \frac{\text{opp}}{\text{adj}} = \frac{x}{1} = x$$

$$\tan 60° = \frac{\text{opp}}{\text{adj}} = \frac{x+y}{1} = x + y$$

4. **Solve for the values of *x* and *y*.**

The tangents of 30-degree and 60-degree angles are convenient values. If you refer to the Appendix, you see that $\tan 30° = \frac{\sqrt{3}}{3} = x$ and $\tan 60° = \sqrt{3} = x + y$.

The value of *y* is the distance that the rocket traveled between the first and second sightings, so, solving for *y*, you get $y = x + y - x = \sqrt{3} - \frac{\sqrt{3}}{3} = \frac{3\sqrt{3}}{3} - \frac{\sqrt{3}}{3} = \frac{2\sqrt{3}}{3} \approx 1.155$. The rocket rose about 1.155 miles in two seconds.

Sine and cosine with algebra

You can fairly accurately approximate the sine and cosine of angles with an *infinite series,* which is the sum of the terms of some sequence, or list, of numbers. Take note, however, that the series for sine and cosine are only accurate for angles from about −90 degrees to 90 degrees. The series for the sine of an angle is $\sin x = x - \frac{x^3}{3!} + \frac{x^5}{5!} - \frac{x^7}{7!} + \frac{x^9}{9!}$, and the series for the cosine of an angle is $\cos x = 1 - \frac{x^2}{2!} + \frac{x^4}{4!} - \frac{x^6}{6!} + \frac{x^8}{8!} - \ldots$. To use these formulas, you have to write the angle measure, *x,* in radians and carry it out several places. The exclamations in the formulas don't mean "Oh, goodness! It's a 3!" The exclamations are mathematical operations called *factorials*. Factorial means to multiply that number times every positive integer smaller than it. Going back to the series for the sine, an angle of 30 degrees is about 0.5236 radians. To find sin 0.5236, use the formula to get

$$\sin 0.5236 = 0.5236 - \frac{(0.5236)^3}{3!} + \frac{(0.5236)^5}{5!} - \frac{(0.5236)^7}{7!}$$
$$= 0.5236 - 0.0239 + 0.000328 - 0.000002$$
$$= 0.500026$$

The result is pretty close to the sine of 30 degrees, which is $\frac{1}{2}$. The closer the angle measure is to 0, the more quickly the value of the sine or the cosine meets the exact value (the fewer terms are necessary for the answer).

Measuring the view of satellite cameras

Consider a satellite that orbits Earth at an altitude of 750 miles. Earth has a radius of 3950 miles. How far in any direction can the satellite's cameras see? Figure 9-10 shows the satellite and the length of the camera's scope due to the curvature of Earth.

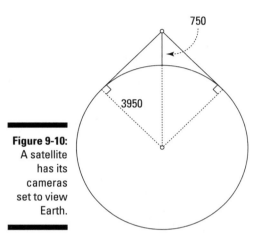

750

3950

Figure 9-10:
A satellite
has its
cameras
set to view
Earth.

1. **Identify the parts of the triangle that you can use to solve the problem.**

   Because a satellite's line of sight is tangent to the curvature of Earth, and tangents to a circle form 90-degree angles with radii of the circle, you can see two right triangles in Figure 9-10. Two radii form angle θ: the one touching the tangent to the circle and the one extending up to the satellite. These radii are the hypotenuse and adjacent side of the right triangle with acute angle θ.

2. **Determine which trig function to use.**

   The adjacent side and hypotenuse are part of the ratio for the cosine of θ.

3. **Write the equation with the trig function, then input the measures that you know and solve for cos θ.**

   The adjacent side measures 3950 miles, and the hypotenuse is the sum of the radius and height of the satellite: 3950 + 750 = 4700 miles.

   $$\cos \theta = \frac{\text{adjacent}}{\text{hypotenuse}} = \frac{3950}{4700} \approx 0.8404$$

4. **Determine the value of θ.**

   Refer to the Appendix to find the angle whose cosine is closest to 0.8404. To the nearer degree, an angle of 33 degrees has this cosine.

5. **Determine how much of Earth's circumference is covered in either direction from the satellite.**

   The satellite's line of sight goes 33 degrees in either direction, or 66 degrees total, which is $\frac{66}{360}$ of the entire circumference (because all the way around would be 360 degrees). If the radius of Earth is 3950 miles, then you can substitute that number into the equation for a circle's circumference: $C = 2\pi r = 2(3.14)(3950) \approx 24,819$. The distance that the satellite scans, then, is $\frac{66}{360} \cdot 24,819 \approx 4550$, or about 4550 miles in any direction.

Calculating Odd Shapes and Maneuvering Corners

Sometimes, finding a measure isn't so easy. You may have to deal with an irregular shape or even calculate your way around a fixed object. Whatever the case, you can use trigonometry to find the answers you've been searching for.

Finding the area of a triangular piece of land

The most commonly used formula for the area of a triangle is $A = \frac{1}{2}bh$, where A is the area, b is the length of the triangle's base, and h is the height of the triangle drawn perpendicular to that base. Figure 9-11 illustrates the different components of this formula.

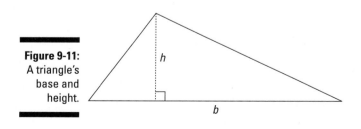

Figure 9-11: A triangle's base and height.

This formula works fine if you can get the measure of the base and the height, and if you can be sure that you've measured a height that's perpendicular to the side of the triangle. But what if you have a triangular yard — a *big* triangular yard — and have no way of measuring some perpendicular segment to one of the sides? One alternative is to use Heron's Formula, which uses the measures of all three sides. The other alternative, of course, is to use trigonometry — or, at least, a formula with an angle measure in it. To measure that angle, you can be very sophisticated and get a surveying apparatus, or if you've got a protractor handy, you can do a decent estimate by extending the sides at an angle for a bit and eyeballing the angle size.

Another formula for finding the area of a triangle is $A = \frac{1}{2} ab \sin \theta$, where a and b are two sides of the triangle and θ is the angle formed between those two sides. You don't need the measure of the third side at all, and you certainly don't need a perpendicular side.

Using trigonometry, I show you where this formula comes from. Take a look at the triangle in Figure 9-12, with sides a and b and the angle between them.

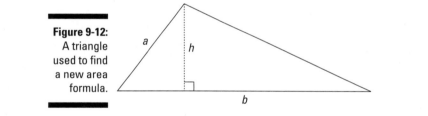

Figure 9-12:
A triangle used to find a new area formula.

Start with the traditional formula for the area of this triangle, $A = \frac{1}{2} bh$. Then look at the little triangle to the left. Because the height is perpendicular to the base, the sides and height form a right triangle. The acute angle θ has a sine equivalent to the following: $\sin \theta = \dfrac{\text{opposite}}{\text{hypotenuse}} = \dfrac{h}{a}$. If you solve that equation for h by multiplying each side by a, you get

$$\sin \theta = \frac{h}{a}$$
$$a \sin \theta = h$$

Replace the h in the traditional formula with its equivalent from the preceding formula, and you get

$$A = \frac{1}{2} bh = \frac{1}{2} b (a \sin \theta) = \frac{1}{2} ab \sin \theta$$

Check out how this formula works in an actual problem. The triangle in Figure 9-13 shows the measures of two of its sides and the angle between them.

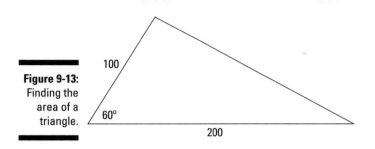

Figure 9-13:
Finding the
area of a
triangle.

100

60°

200

To find the area of the triangle in Figure 9-13:

1. **Use the formula** $A = \frac{1}{2}\,ab\sin\theta$, **inserting the values that you know.**

   $A = \frac{1}{2}(100)(200)\sin 60°$

2. **Solve for the value of the area.**

   $A = \frac{1}{2}(100)(200) \cdot \frac{\sqrt{3}}{2}$

   $A = 10,000 \cdot \frac{\sqrt{3}}{2} \approx 8660$

   The area is about 8660 square units.

Moving an object around a corner

Here's an application of trigonometry that you may very well be able to relate to: Have you ever tried to get a large piece of furniture around a corner in a house? You twist and turn and put it up on end, but to no avail. In this example, pretend that you're trying to get a 15-foot ladder around a corner where two 4-foot-wide hallways meet at a 90-degree angle. Figure 9-14 shows a picture of the situation.

The tightest part comes when the ladder is halfway through the hallway, or when the angles where it touches the outer walls are the same. When the ladder is at the tightest point, it'll form a right triangle with equal sides — half the ladder to each side of the corner. Because the sides of the right triangle are equal at this point, you've got an isosceles right triangle, which has two 45-degree angles (see Figure 9-15, which shows the longest a ladder can be to fit around the corner). How long are the sides of the right triangle, then? When you know the dimensions of this isosceles right triangle, then you can look at the hypotenuse — the ladder — and determine if it's short enough or too long to fit around the tightest part of this corner. And, of course, you don't want to scrape or punch holes in the wall!

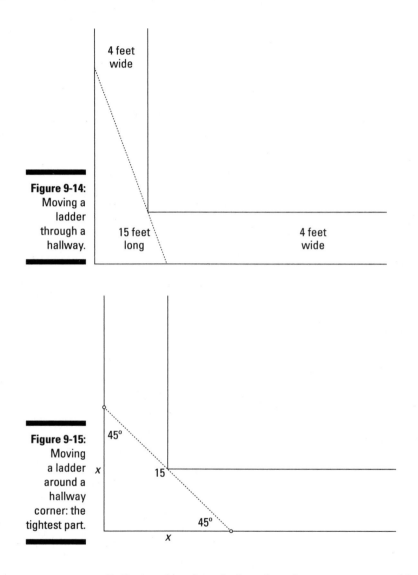

Figure 9-14:
Moving a
ladder
through a
hallway.

Figure 9-15:
Moving
a ladder
around a
hallway
corner: the
tightest part.

1. **Determine the trig function that you can use with the measures available.**

   The hypotenuse is the length of the ladder — 15 feet. The opposite and adjacent sides are the same in an isosceles right triangle, and in this case, those two lengths are each 8 feet. You know this measure because all the triangles are isosceles right triangles, which means they have 45-degree angles and equal leg measures (see Figure 9-16).

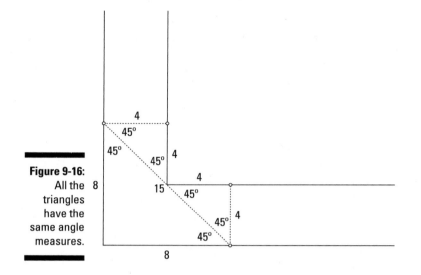

2. **Determine which trig function to use.**

   Both sine and cosine include the length of the hypotenuse, which is what you're solving for, so you can use either function.

3. **Write the equation with the trig function, then input the measures that you know.**

   $$\sin 45° = \frac{\text{opposite}}{\text{hypotenuse}}$$
   $$\sin 45° = \frac{8}{\text{hypotenuse}}$$

4. **Solve for the value of the hypotenuse.**

   $$\frac{\sqrt{2}}{2} = \frac{8}{\text{hypotenuse}}$$
   $$\sqrt{2} \cdot \text{hypotenuse} = 8 \cdot 2$$
   $$\text{hypotenuse} = \frac{16}{\sqrt{2}} \approx 11.314$$

   You find that at the tightest point around the corner, the hypotenuse is only slightly more than 11 feet. That 15-foot ladder will never fit around the corner.

Part III
Identities

The 5th Wave   By Rich Tennant

$$\left(xA-xB\right)^2 + \left(yA-yB\right)^2$$

"...and so, with this formula, Descartes was able to compute the distance in a plane. However, despite this accomplishment, his luggage was still as likely to get lost on route as it had before."

In this part . . .

Trig identities are enlightening, yet mysterious. They're helpful, yet frustrating. But most of all, they're absolutely necessary. I start off by showing you the basic identities, and then I move on to the more-complex ones. And just to prove that I didn't pull these identities out of thin air, I use algebra techniques to solve and confirm the identities.

Chapter 10

Trig's Basic Identities

I'm sure a thousand questions are running through your mind: What's a trig identity? Is it possible to have a mistaken trig identity? Does anyone commit trig-identity theft? Can you have an identity crisis with trig? The answer: no.

Trig identities aren't nearly so sinister. They're actually very helpful tools in simplifying trig expressions and solving equations. These identities are special to trigonometry. Basically, they're equivalences — they give you options to substitute into equations in order to simplify.

Identities are divided into different types, or categories, in order to help you remember them more easily and figure out when to use them more efficiently; in this chapter, I cover the gamut.

Flipping Functions on Their Backs: Reciprocal Identities

The simplest and most basic trig identities are those involving the reciprocals of the trig functions. To jog your memory, a *reciprocal* of a number is 1 divided by that number — for example, the reciprocal of 2 is $\frac{1}{2}$. Another way to describe reciprocals is to point out that the product of a number and its reciprocal is 1. In the case of 2 and its reciprocal, $\frac{1}{2}$, $2 \cdot \frac{1}{2} = 1$. The same goes for the trig reciprocals.

A woman ahead of her time: Hypatia

One of the earliest recognized female mathematicians was Hypatia, who lived in Alexandria, Egypt, and is thought to have been born around A.D. 370. People of her day considered Hypatia to be not only a mathematician, but also a scientist and philosopher. Her father, Theon, was a professor of mathematics at the University of Alexandria. He taught Hypatia himself and shared with her his passion for knowledge and the search for answers. Hypatia developed a great enthusiasm for mathematics as well as astronomy and astrology. Her father also believed in a strong and healthy body as well as mind, so he insisted on a regular physical routine to achieve this standard of excellence.

Hypatia is well-known for her work on the ideas of conic sections, which are the curves that are formed by slicing a cone in various ways. The curves are called *parabola, circle, ellipse,* and *hyperbola.* She edited the work of Apollonius, making the concepts easier to understand. For this, she is considered to be the first woman to make a contribution resulting in the survival of some of the earlier mathematical ideas.

Hypatia came to a tragic end, killed by a mob who were spurred on by rumors created by leaders who didn't appreciate her religious stand or alliances.

Here's how the reciprocal identities are defined:

- The reciprocal of sine is cosecant: $\dfrac{1}{\sin\theta} = \csc\theta$

- The reciprocal of cosine is secant: $\dfrac{1}{\cos\theta} = \sec\theta$

- The reciprocal of tangent is cotangent: $\dfrac{1}{\tan\theta} = \cot\theta$

- The reciprocal of cotangent is tangent: $\dfrac{1}{\cot\theta} = \tan\theta$

- The reciprocal of secant is cosine: $\dfrac{1}{\sec\theta} = \cos\theta$

- The reciprocal of cosecant is sine: $\dfrac{1}{\csc\theta} = \sin\theta$

In true fashion, when you multiply the reciprocals together, you get 1:

$$\sin\theta \cdot \csc\theta = 1$$

$$\cos\theta \cdot \sec\theta = 1$$

$$\tan\theta \cdot \cot\theta = 1$$

The reciprocal identity is a very useful one when you're solving trig equations. If you find a way to multiply each side of an equation by a function's reciprocal, you may be able to reduce some part of the equation to 1 — and simplifying is always a good thing.

Function to Function: Ratio Identities

Trig has two identities called *ratio identities.* This label can be confusing, because all the trig functions are defined by ratios. Somewhere along the line, however, mathematicians thought this description was perfect for these two identities, because they're basically fractions made up of two trig functions, one above the other, in each. The ratio identities create ways to write tangent and cotangent by using the other two basic functions, sine and cosine.

The ratio identities are $\tan\theta = \dfrac{\sin\theta}{\cos\theta}$ and $\cot\theta = \dfrac{\cos\theta}{\sin\theta}$.

These two identities come from the simplification of a couple of complex fractions. If you use the basic definitions for sine, cosine, and tangent, $\text{sine} = \dfrac{\text{opp}}{\text{hyp}}$, $\text{cosine} = \dfrac{\text{adj}}{\text{hyp}}$, and $\text{tangent} = \dfrac{\text{opp}}{\text{adj}}$, then you can see that

$$\frac{\text{sine}}{\text{cosine}} = \frac{\dfrac{\text{opp}}{\text{hyp}}}{\dfrac{\text{adj}}{\text{hyp}}} = \frac{\text{opp}}{\cancel{\text{hyp}}} \cdot \frac{\cancel{\text{hyp}}}{\text{adj}} = \frac{\text{opp}}{\text{adj}} = \text{tangent}$$

Likewise, because cotangent is the reciprocal of tangent,

$$\text{cotangent} = \frac{1}{\text{tangent}} = \frac{1}{\dfrac{\text{sine}}{\text{cosine}}} = \frac{\text{cosine}}{\text{sine}}$$

Opposites Attract: Opposite-Angle Identities

The *opposite-angle identities* change functions of negative angles to functions of positive angles. Negative angles are great for describing a situation, but they aren't really handy when it comes to sticking them in a trig function and calculating that value. So, for example, you can rewrite the sine of −30 degrees as the sine of 30 degrees by putting a negative sign in front of the function:

$$\sin(-30°) = -\sin(30°)$$

The identity works differently for different functions, though. First, consider the identities, and then find out how they came to be.

The opposite-angle identities for the three most basic functions are

$$\sin(-\theta) = -\sin\theta$$
$$\cos(-\theta) = \cos\theta$$
$$\tan(-\theta) = -\tan\theta$$

The rule for the sine and tangent of a negative angle almost seems intuitive. But what's with the cosine? How can the cosine of a negative angle be the same as the cosine of the corresponding positive angle? Here's how it works.

If you refer back to Chapter 8, you find out that the functions of angles with their terminal sides in the different quadrants have varying signs. Sine, for example, is positive when the angle's terminal side lies in the first and second quadrants, whereas cosine is positive in the first and fourth quadrants. In addition, Chapter 4 shows you how to draw angles on a coordinate plane: Positive angles go counterclockwise from the positive x-axis, and negative angles go clockwise.

With those points in mind, take a look at Figure 10-1, which shows a –45-degree angle and a 45-degree angle.

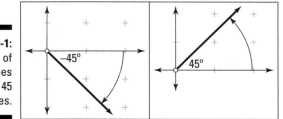

Figure 10-1:
Angles of
–45 degrees
and 45
degrees.

First, consider the –45-degree angle. This angle has its terminal side in the fourth quadrant, so its sine is negative. A 45-degree angle, on the other hand, has a positive sine, so $\sin(-45°) = -\sin(45°) = -\frac{\sqrt{2}}{2}$. In plain English, the sine of a negative angle is the opposite value of that of the positive angle with the same measure.

A new way to slice pi-e

Two of the best-known symbols used for two constants in mathematics are the Greek letter pi, π, and the lowercase letter e. The value of π, approximately 3.14159, is a decimal that goes on forever. The value of e, approximately 2.71828, goes on forever, too.

The symbol for pi was first introduced in the early 1700s by William Jones, an obscure English writer who was composing a book for math beginners. Common belief is that he chose this letter because p is the first letter in *perimetron*, meaning perimeter. He didn't realize that what he was doing would have such a long-lasting effect on the world of mathematics. A few years later, the famous mathematician Leonhard Euler used the letter p instead of the Greek letter π, but he eventually adopted the Greek notation, too, making it even more popular with the rest of the world.

The letter e represents the base for natural logarithms. Like π, this value occurs naturally — in other words, the value e occurs as a multiplier or base of the equations that represent many natural phenomena. Euler also had a hand in popularizing this symbol sometime in the early to mid-1700s. The question comes to mind as to why he chose that particular letter. Using π or p seems natural for a concept linked to the perimeter of a circle, but the letter e isn't such an obvious choice. Here are a few possible reasons why our ancestors chose this letter: First, e is the first letter in *exponential*, which is closely tied to logarithms; or perhaps they used e because it's near the beginning of the alphabet and hadn't been used as a symbol for anything else. Could it be, rather, that Euler chose the letter e because it's the first letter in his name? Whatever the reason, it appears that π and e are here to stay.

Now on to the cosine function. In light of the cosine's sign with respect to the coordinate plane, you know that an angle of –45 degrees has a positive cosine. So does its counterpart, the angle of 45 degrees, which is why cos (–45°) = $\cos(45°) = \dfrac{\sqrt{2}}{2}$. So you see, the cosine of a negative angle is the same as that of the positive angle with the same measure.

Next, try the identity on another angle, a negative angle with its terminal side in the third quadrant. Figure 10-2 shows a negative angle with the measure of –120 degrees and its corresponding positive angle, 120 degrees.

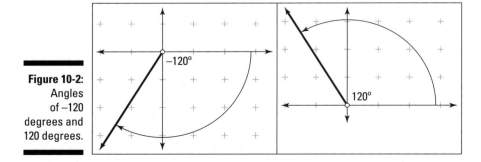

Figure 10-2:
Angles
of –120
degrees and
120 degrees.

The angle of –120 degrees has its terminal side in the third quadrant, so both its sine and cosine are negative. Its counterpart, the angle measuring 120 degrees, has its terminal side in the second quadrant, where the sine is positive and the cosine is negative. So the sine of –120 degrees is the opposite of the sine of 120 degrees, and the cosine of –120 degrees is the same as the cosine of 120 degrees. In trig notation, it looks like this:

$$\sin(-120°) = -\sin(120°) = -\frac{\sqrt{3}}{2} \text{ and } \cos(-120°) = \cos(120°) = -\frac{1}{2}$$

When you apply the opposite-angle identity to the tangent of a 120-degree angle (which comes out to be negative), you get that the opposite of a negative is a positive. Surprise, surprise. So, applying the identity, the opposite makes the tangent positive, which is what you get when you take the tangent of 120 degrees, where the terminal side is in the third quadrant and is therefore positive.

Revisiting the Classic Theorem: Pythagorean Identities

Good old Pythagoras is at work everywhere — his theorem keeps cropping up in the strangest places. (Not that a chapter in this book is really a strange place, of course.) For more on that theorem, refer to Chapter 2. This section, however, takes you past the basics and expands on them with the three identities called *Pythagorean identities*.

The Pythagorean identities are building blocks for many of the manipulations of equations and expressions. They provide a greater number of methods for solving problems more efficiently, because they allow you to write complicated expressions in a much simpler form.

The Pythagorean identities are

$$\sin^2\theta + \cos^2\theta = 1$$
$$\tan^2\theta + 1 = \sec^2\theta$$
$$1 + \cot^2\theta = \csc^2\theta$$

The notation in these identities is peculiar to trigonometry. The expression $\sin^2\theta$ actually means $(\sin\theta)^2$: Find the sine of angle θ and then square that number. But mathematicians hate to waste, and having to put those big, cumbersome parentheses around "sin θ" all the time seemed wasteful. So they agreed on a condensed version: The superscript 2 right after "sin" means that you square the whole expression. The same type of notation also goes for the other functions.

The mother of all Pythagorean identities

The Pythagorean identity that birthed the other two is $\sin^2\theta + \cos^2\theta = 1$. But, you may wonder, where did this identity come from, and why is it so important? Last things first: The primary identity is important because it sets a combination of functions equal to 1, and this simplification is very helpful for solving trig equations. As such, it's probably one of the most frequently used identities of them all. This identity comes from putting a right triangle inside the unit circle and substituting values and equations to come up with a whole new equation (see Figure 10-3).

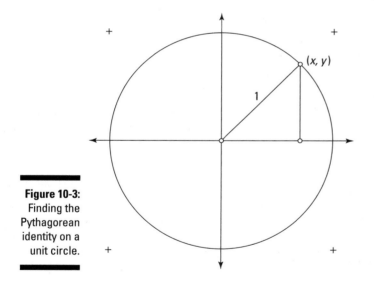

Figure 10-3:
Finding the
Pythagorean
identity on a
unit circle.

In Chapter 8, you discover that in a circle, $\sin\theta = \frac{y}{r}$ and $\cos\theta = \frac{x}{r}$, where (x, y) are the coordinates of the point and r is the radius of the circle. The value of x is also the length of the adjacent side of the triangle, and y is the length of the opposite side. In a unit circle, the radius is equal to 1. When you substitute that value in the equation, you find that $\sin\theta = \frac{y}{1} = y$ and $\cos\theta = \frac{x}{1} = x$. Hold that thought.

The Pythagorean theorem says that when you square the value of each of a triangle's two legs and add the results together, you get the square of the hypotenuse. In mathematical notation, it looks like this: $a^2 + b^2 = c^2$. In the case of the right triangle on the unit circle, because the radius (which is also the hypotenuse) is 1, you can say that $x^2 + y^2 = 1^2$. Now replace the x with $\cos\theta$ and the y with $\sin\theta$, switch the two terms around, and you get $\sin^2\theta + \cos^2\theta = 1$.

If all the finagling just seems like a lot of hocus pocus to you, check out this identity in action. Suppose the angle in question is 30 degrees. Using the values for the functions of a 30-degree angle (see the Appendix), $\sin 30° = \frac{1}{2}$ and $\cos 30° = \frac{\sqrt{3}}{2}$, and putting them into the identity, you get

$$\sin^2\theta + \cos^2\theta = 1$$

$$\left(\frac{1}{2}\right)^2 + \left(\frac{\sqrt{3}}{2}\right)^2 = \frac{1}{4} + \frac{3}{4} = 1$$

Voilà!

Extending to tangent and secant

The other two Pythagorean identities stem from the first one. All you do is throw in a little algebra and apply the reciprocal and ratio identities (see those sections earlier in this chapter) and — poof! — two new identities.

1. **Starting with the first Pythagorean identity, $\sin^2\theta + \cos^2\theta = 1$, divide each term by $\cos^2\theta$.**

    $$\frac{\sin^2\theta}{\cos^2\theta} + \frac{\cos^2\theta}{\cos^2\theta} = \frac{1}{\cos^2\theta}$$

2. **Rewrite each term by using the algebraic rule $\left(\frac{a^2}{b^2}\right) = \left(\frac{a}{b}\right)^2$.**

    Note how the exponent 2 is pulled out of the parentheses:

    $$\left(\frac{\sin\theta}{\cos\theta}\right)^2 + \left(\frac{\cos\theta}{\cos\theta}\right)^2 = \left(\frac{1}{\cos\theta}\right)^2$$

3. **Replace each of the terms with an equivalent expression.**

    - Use a ratio identity, $\frac{\sin\theta}{\cos\theta} = \tan\theta$, to replace the first term with $\tan\theta$.

    - Replace the second term with 1: $\frac{\cos\theta}{\cos\theta} = 1$.

    - Use a reciprocal identity, $\frac{1}{\cos\theta} = \sec\theta$, to replace the third term with $\sec\theta$.

Substituting these expressions into the equation and simplifying, you get

$$(\tan\theta)^2 + (1)^2 = (\sec\theta)^2$$
$$\tan^2\theta + 1 = \sec^2\theta$$

Finishing up with cotangent and cosecant

If you read the preceding section, you can simply do a repeat performance with cotangent and cosecant. Here's how:

1. **Starting with the first Pythagorean identity, $\sin^2\theta + \cos^2\theta = 1$, divide each term by $\sin^2\theta$.**

$$\frac{\sin^2\theta}{\sin^2\theta} + \frac{\cos^2\theta}{\sin^2\theta} = \frac{1}{\sin^2\theta}$$

2. **Rewrite each term by using the algebraic rule $\left(\dfrac{a^2}{b^2}\right) = \left(\dfrac{a}{b}\right)^2$.**

$$\left(\frac{\sin\theta}{\sin\theta}\right)^2 + \left(\frac{\cos\theta}{\sin\theta}\right)^2 = \left(\frac{1}{\sin\theta}\right)^2$$

3. **Replace each of the terms with an equivalent expression.**

   - Replace the first term with 1: $\dfrac{\sin\theta}{\sin\theta} = 1$.
   - Use a ratio identity, $\dfrac{\cos\theta}{\sin\theta} = \cot\theta$, to replace the second term with $\cot\theta$.
   - Use a reciprocal identity, $\dfrac{1}{\sin\theta} = \csc\theta$, to replace the third term with $\csc\theta$.

   Substituting these expressions into the equation and simplifying, you find that the result is

$$(1)^2 + (\cot\theta)^2 = (\csc\theta)^2$$
$$1 + \cot^2\theta = \csc^2\theta$$

Rearranging the Pythagorean identities

The preceding sections show the original definitions of the Pythagorean identities, but as you probably suspected, the forms don't end there. Familiarizing yourself with the other versions of these identities is helpful so that you can easily recognize them when solving equations or simplifying expressions.

All these different versions have their places in trigonometric applications, calculus, or other math topics. You don't have to memorize them, because if you just remember the three Pythagorean identities, you can solve for what you need.

John Napier, inventor of logs and bones

John Napier was a 16th-century Scottish mathematician and inventor credited with inventing logarithms (for which he is best known), the decimal point, and *Napier's Bones* — an early calculating instrument. Napier's Bones were actually strips of wood or bone with multiplication tables inscribed on them. People used them for multiplying, dividing, taking square and cube roots, determining decimal values of fractions, and doing computations with exponential and trig functions.

As an inventor, Napier created a hydraulic screw, which was used in coal pits to lower the water levels. He invented, or at least proposed, several military inventions as well. He proposed special artillery; bulletproof clothing; a submarinelike vehicle; and huge, burning mirrors to set enemy ships on fire.

Changing $\sin^2\theta + \cos^2\theta = 1$

You can alter the original Pythagorean identity in myriad ways. For starters, you can isolate either $\sin^2\theta$ or $\cos^2\theta$ on one side of the equation by subtracting the other term:

$$\sin^2\theta = 1 - \cos^2\theta$$
$$\cos^2\theta = 1 - \sin^2\theta$$

Continuing on, you can factor the right side of either of these equations because that side is the difference of two perfect squares:

$$\sin^2\theta = 1 - \cos^2\theta = (1 - \cos\theta)(1 + \cos\theta)$$
$$\cos^2\theta = 1 - \sin^2\theta = (1 - \sin\theta)(1 + \sin\theta)$$

Sometimes, however, having an expression for $\sin\theta$ or $\cos\theta$, where the functions aren't squared, is helpful. Beginning with the earlier version of the basic Pythagorean identities, where one function is by itself, you can take the square root of each side to get

$$\sin\theta = \pm\sqrt{1 - \cos^2\theta} \text{ or } \cos\theta = \pm\sqrt{1 - \sin^2\theta}$$

Adjusting $\tan^2\theta + 1 = \sec^2\theta$

You can also adapt this second Pythagorean identity in various ways. Solving for $\tan^2\theta$ by subtracting 1 from each side of the equation, you get

$$\tan^2\theta = \sec^2\theta - 1$$

Then, factoring the difference of the squares on the right (because that side is the difference of two perfect squares), you have

$$\tan^2\theta = (\sec\theta - 1)(\sec\theta + 1)$$

Lastly, beginning with the earlier version and taking the square root of each side, you get

$$\tan\theta = \pm\sqrt{\sec^2\theta - 1}$$

Taking another approach to this Pythagorean identity, you can subtract $\tan^2\theta$ from each side and factor the result to get

$$1 = \sec^2\theta - \tan^2\theta = (\sec\theta - \tan\theta)(\sec\theta + \tan\theta)$$

Reconfiguring $1 + \cot^2\theta = \csc^2\theta$

You can rearrange the last Pythagorean identity, too, by subtracting 1 from each side or by subtracting $\cot^2\theta$ from each side. The two new versions are

$$\cot^2\theta = \csc^2\theta - 1$$
$$1 = \csc^2\theta - \cot^2\theta$$

Each of the preceding equations has the difference of two perfect squares, which you can factor:

$$\cot^2\theta = \csc^2\theta - 1 = (\csc\theta - 1)(\csc\theta + 1)$$
$$1 = \csc^2\theta - \cot^2\theta = (\csc\theta - \cot\theta)(\csc\theta + \cot\theta)$$

And lastly, the square root of each side yields an identity involving just $\cot\theta$:

$$\cot\theta = \pm\sqrt{\csc^2\theta - 1}$$

Combining the Identities

Even though each function is perfectly wonderful, being able to express them all in terms of one of the other five trig functions is frequently to your advantage. For example, you may have some other terms in an expression that aren't sines, and having them all match would help you solve the equation.

Armed with the reciprocal identities, ratio identities, and Pythagorean identities, you can do just that — write any trig function in terms of the others. In this section, I show you the many variations of sine; by applying some of the same identities and following similar steps, you can form a multitude of variations of the other trig functions. These five ways of writing sine in terms of the other five functions show you how powerful, versatile, and useful all these identities are. A word of warning: Some of these equations aren't very pretty.

The many faces of sine

Here are the five ways of expressing the sine function in terms of the other functions:

Sine in terms of cosine: $\sin\theta = \pm\sqrt{1 - \cos^2\theta}$

Sine in terms of tangent: $\sin\theta = \dfrac{\tan\theta}{\pm\sqrt{\tan^2\theta + 1}}$

Sine in terms of cotangent: $\sin\theta = \dfrac{1}{\pm\sqrt{1 + \cot^2\theta}}$

Sine in terms of secant: $\sin\theta = \dfrac{\pm\sqrt{\sec\theta - 1}}{\sec\theta}$

Sine in terms of cosecant: $\sin\theta = \dfrac{1}{\csc\theta}$

Working out the versions

Choosing a good starting point helps — and makes for a nicer result. You don't want any expression that's too messy or hard to remember. Take advantage of identities that have your target function isolated in a single term. This section shows you the most typical methods for changing one trig function to another.

Changing sine to cosine

You can express the sine function in terms of cosine without doing much work. Starting with the Pythagorean identity involving $\sin\theta$ and $\cos\theta$, subtract $\cos^2\theta$ from each side and take the square root of each side:

$$\sin^2\theta = 1 - \cos^2\theta$$
$$\sqrt{\sin^2\theta} = \pm\sqrt{1 - \cos^2\theta}$$
$$\sin\theta = \pm\sqrt{1 - \cos^2\theta}$$

Changing sine to tangent

To rewrite the sine function in terms of tangent:

1. **Start with the ratio identity involving sine, cosine, and tangent, and multiply each side by cosine to get the sine alone on the left.**

$$\frac{\sin\theta}{\cos\theta} = \tan\theta$$
$$\sin\theta = \tan\theta\cos\theta$$

2. **Replace cosine with its reciprocal function.**

$$\sin \theta = \tan \theta \left(\frac{1}{\sec \theta} \right)$$

3. **Solve the Pythagorean identity $\tan^2 \theta + 1 = \sec^2 \theta$ for secant.**

   This equation gives you $\pm \sqrt{\tan^2 \theta + 1} = \sec \theta$.

4. **Replace the secant in the sine equation.**

   You end up with $\sin \theta = \tan \theta \left(\dfrac{1}{\pm \sqrt{\tan^2 \theta + 1}} \right) = \dfrac{\tan \theta}{\pm \sqrt{\tan^2 \theta + 1}}$.

Changing sine to cotangent

To write the sine function in terms of cotangent, begin with the equation you end up with in the preceding section.

1. **Replace all the tangents with 1 over the reciprocal for tangent (which is cotangent) and simplify the expression.**

$$\sin \theta = \frac{\dfrac{1}{\cot \theta}}{\pm \sqrt{\dfrac{1}{\cot^2 \theta} + 1}}$$

   The result is a *complex fraction* — it has fractions in both the numerator and denominator — so it'll look a lot better if you simplify it.

2. **Rewrite the part under the radical as a single fraction and simplify it by taking the square root of each part.**

$$\sin \theta \frac{\dfrac{1}{\cot \theta}}{\pm \sqrt{\dfrac{1}{\cot^2 \theta} + \dfrac{\cot^2 \theta}{\cot^2 \theta}}} = \frac{\dfrac{1}{\cot \theta}}{\pm \sqrt{\dfrac{1 + \cot^2 \theta}{\cot^2 \theta}}}$$

$$= \frac{\dfrac{1}{\cot \theta}}{\pm \dfrac{\sqrt{1 + \cot^2 \theta}}{\sqrt{\cot^2 \theta}}} = \frac{\dfrac{1}{\cot \theta}}{\pm \dfrac{\sqrt{1 + \cot^2 \theta}}{\cot \theta}}$$

3. **Multiply the numerator by the reciprocal of the denominator.**

$$= \frac{\dfrac{1}{\cot \theta}}{\pm \dfrac{\sqrt{1 + \cot^2 \theta}}{\cot \theta}} = \frac{1}{\cancel{\cot \theta}} \cdot \frac{\cancel{\cot \theta}}{\pm \sqrt{1 + \cot^2 \theta}} = \frac{1}{\pm \sqrt{1 + \cot^2 \theta}}$$

   Voilà — you have sine in terms of cotangent.

Changing sine to secant

The next function to define sine is the secant function. The easiest way to do this is to start with the sine in terms of cosine and replace each cosine with 1 over its reciprocal (which is secant):

$$\sin\theta = \pm\sqrt{1 - \cos^2\theta}$$
$$= \pm\sqrt{1 - \frac{1}{\sec^2\theta}}$$

The radical has a fraction in it. A better form is to simplify that fraction, so find a common denominator and split the fraction into two radicals — the bottom one of which you can further simplify:

$$\sin\theta = \pm\sqrt{\frac{\sec^2\theta}{\sec^2\theta} - \frac{1}{\sec^2\theta}} = \pm\sqrt{\frac{\sec^2\theta - 1}{\sec^2\theta}} = \pm\frac{\sqrt{\sec^2\theta - 1}}{\sec\theta}$$

Changing sine to cosecant

The last function to write sine in terms of is the cosecant — I saved the best (easiest) for last. The reciprocal of cosecant is sine, so this equation is just one of the basic reciprocal identities:

$$\sin\theta = \frac{1}{\csc\theta}$$

Chapter 11

Operating on Identities

*T*he basic building-block identities are the reciprocal, ratio, and Pythagorean identities, which I discuss in detail in Chapter 10. In this chapter, you take those identities a step further and develop new identities, discovering how to add, subtract, multiply, and divide the trig functions — in particular, the nice values for angles of 0, 30, 45, 60, and 90 degrees. (Those angles aren't the only ones that you can perform operations on; they're just the most convenient to use when showing how the trig identities work.) By performing such operations, you can determine the function values of even more angles.

Summing It Up

Three basic identities involve the sums of angles, and they involve sine, cosine, and tangent. After you recognize these three identities, you can adapt them for the other three functions (cosecant, secant, and cotangent) by using the reciprocal identities (detailed in Chapter 10). All you do is use a reciprocal identity to change the expression to one of the three basic functions, do the necessary work, and then use the reciprocal identity to change the answer back in terms of the function you began with.

You can use the angle-sum identities to find the function values of any angles, but the examples in this section just show the most convenient combinations — ones with exact values that you can fill into the formulas. Suppose, for example, that you want to find the exact value of the sine of 75 degrees. To minimize fuss, you can use the sum of 30 degrees and 45 degrees and the appropriate identity.

The angle-sum identities find the function value for the sum of angle α and angle β:

$$\sin(\alpha + \beta) = \sin \alpha \cos \beta + \cos \alpha \sin \beta$$

$$\cos(\alpha + \beta) = \cos \alpha \cos \beta - \sin \alpha \sin \beta$$

$$\tan(\alpha + \beta) = \frac{\tan \alpha + \tan \beta}{1 - \tan \alpha \tan \beta}$$

Using the identity for the sine of a sum, find the sine of 75 degrees:

1. **Determine two angles whose sum is 75 for which you know the values for both sine and cosine.**

   Choose 30 + 45, not 50 + 25 or 70 + 5, because sticking to the more-common angles that have nice, exact values to use in the formula is your best bet.

2. **Input the angle measures into the identity.**

   $$\sin(\alpha + \beta) = \sin \alpha \cos \beta + \cos \alpha \sin \beta$$

   $$\sin(30° + 45°) = \sin 30° \cos 45° + \cos 30° \sin 45°$$

3. **Replace the functions of the angles with their values and simplify.**

   $$\sin(30° + 45°) = \sin 30° \cos 45° + \cos 30° \sin 45°$$

   $$= \frac{1}{2} \cdot \frac{\sqrt{2}}{2} + \frac{\sqrt{3}}{2} \cdot \frac{\sqrt{2}}{2}$$

   $$= \frac{\sqrt{2}}{4} + \frac{\sqrt{6}}{4} = \frac{\sqrt{2} + \sqrt{6}}{4} = \sin 75°$$

Sometimes, you have more than one choice for the sum. In this example, find the cosine of 120 degrees by using the identity for the cosine of a sum.

1. **Determine two angles whose sum is 120.**

   Choosing among the most convenient angles, you can use either 90 + 30 or 60 + 60. For this example, I use 90 + 30, because the sine of a 90-degree angle is 1, and the cosine is equal to 0. Both of those numbers are very nice to have in a computation because they keep it simple.

2. **Input the values into the identity.**

   $$\cos(\alpha + \beta) = \cos \alpha \cos \beta - \sin \alpha \sin \beta$$

   $$\cos(90° + 30°) = \cos 90° \cos 30° - \sin 90° \sin 30°$$

3. **Replace the functions with their values and simplify.**

   $$\cos(90° + 30°) = \cos 90° \cos 30° - \sin 90° \sin 30°$$

   $$= 0 \cdot \frac{\sqrt{3}}{2} - 1 \cdot \frac{1}{2} = 0 - \frac{1}{2} = -\frac{1}{2} = \cos(120°)$$

These identities work with radian measures, too, such as finding $\tan \frac{7\pi}{12}$ by using the identity for the tangent of the sum of angles.

1. Determine two angles whose sum is $\frac{7\pi}{12}$.

It may be easier to think of finding two numbers that add up to $\frac{7}{12}$. Simplified, they're $\frac{1}{3}$ and $\frac{1}{4}$, so $\frac{7\pi}{12} = \frac{\pi}{3} + \frac{\pi}{4}$.

2. Input the values into the identity.

$$\tan(\alpha + \beta) = \frac{\tan \alpha + \tan \beta}{1 - \tan \alpha \tan \beta}$$

$$\tan\left(\frac{\pi}{3} + \frac{\pi}{4}\right) = \frac{\tan \frac{\pi}{3} + \tan \frac{\pi}{4}}{1 - \tan \frac{\pi}{3} \tan \frac{\pi}{4}}$$

3. Replace the functions with their values and simplify.

$$\tan\left(\frac{\pi}{3} + \frac{\pi}{4}\right) = \frac{\sqrt{3} + 1}{1 - \sqrt{3} \cdot 1} = \frac{\sqrt{3} + 1}{1 - \sqrt{3}} = \tan \frac{7\pi}{12}$$

The result in the last step doesn't leave the answer in the nicest form. The denominator has two terms, and one of them is a radical. One way to make the answer look a bit better and more intelligible is to use a technique called *rationalization*.

To rationalize the numerator or denominator of a fraction, multiply both the numerator and denominator by the *conjugate* (same terms, opposite sign) of the part that you're rationalizing. When you do so, you end up with the difference of two squares and get rid of the offending portion.

For the last example, you rationalize like so to get the radical out of the denominator:

$$\frac{\sqrt{3} + 1}{1 - \sqrt{3}} \cdot \frac{1 + \sqrt{3}}{1 + \sqrt{3}} = \frac{\sqrt{3} + 3 + 1 + \sqrt{3}}{1 - 3} = \frac{4 + 2\sqrt{3}}{-2} = -2 - \sqrt{3}$$

The final answer is a bit nicer to understand and estimate. Because $\sqrt{3}$ is about 1.7, you can estimate that $-2 - \sqrt{3} \approx -2 - 1.7 = -3.7$.

Sometimes, you may not know what the angle measure is, but you know something about the angle's function values. For example, suppose you have two angles, α in QII and β in QI. If $\sin \alpha = \frac{3}{5}$ and $\cos \beta = \frac{24}{25}$, then what are $\sin(\alpha + \beta)$ and $\cos(\alpha + \beta)$?

1. **Find all the necessary function values for the sums.**

   Both the sine and cosine angle-sum identities use the sine and cosine of each angle involved. You already know the sine of one angle and the cosine of the other angle, so you have to determine the unknown cosine and sine — you can do so by using the Pythagorean identity:

   - First, use the value for $\sin \alpha$ to solve for $\cos \alpha$:

     $$\sin^2 \alpha + \cos^2 \alpha = 1$$

     $$\left(\frac{3}{5}\right)^2 + \cos^2 \alpha = 1$$

     $$\cos^2 \alpha = 1 - \frac{9}{25} = \frac{16}{25}$$

     $$\cos \alpha = \pm \sqrt{\frac{16}{25}} = \pm \frac{4}{5}$$

     You end up with two results. Because the terminal side of angle α is in the second quadrant, the cosine of α, in this case, is negative:
     $$\cos \alpha = -\frac{4}{5}.$$

   - Now use the value for $\cos \beta$ to solve for $\sin \beta$:

     $$\sin^2 \beta + \cos^2 \beta = 1$$

     $$\sin^2 \beta + \left(\frac{24}{25}\right)^2 = 1$$

     $$\sin^2 \beta = 1 - \frac{576}{625} = \frac{625}{49}$$

     $$\sin \beta = \pm \sqrt{\frac{49}{625}} = \pm \frac{7}{25}$$

     The terminal side of angle β is in the first quadrant, where the sine is positive: $\sin \beta = \frac{7}{25}$.

2. **Insert the function values into the identities for the sine and cosine of the sum of angles.**

   $$\sin(\alpha + \beta) = \sin \alpha \cos \beta + \cos \alpha \sin \beta$$

   $$\sin(\alpha + \beta) = \left(\frac{3}{5}\right)\left(\frac{24}{25}\right) + \left(-\frac{4}{5}\right)\left(\frac{7}{25}\right)$$

   $$\cos(\alpha + \beta) = \cos \alpha \cos \beta - \sin \alpha \sin \beta$$

   $$\cos(\alpha + \beta) = \left(-\frac{4}{5}\right)\left(\frac{24}{25}\right) - \left(\frac{3}{5}\right)\left(\frac{7}{25}\right)$$

3. **Simplify the identities and solve for the answers.**

   $$\sin(\alpha + \beta) = \left(\frac{3}{5}\right)\left(\frac{24}{25}\right) + \left(-\frac{4}{5}\right)\left(\frac{7}{25}\right)$$

   $$= \left(\frac{72}{125}\right) + \left(-\frac{28}{125}\right) = \frac{44}{125}$$

$$\cos(\alpha + \beta) = \left(-\frac{4}{5}\right)\left(\frac{24}{25}\right) - \left(\frac{3}{5}\right)\left(\frac{7}{25}\right)$$
$$= \left(-\frac{96}{125}\right) - \left(\frac{21}{125}\right) = -\frac{117}{125}$$

By looking at the angle measures, you can predict whether the function value will be positive or negative. In the preceding example, the smaller angles, when added together, create an angle with its terminal side in the second quadrant. In Chapter 8 you find out that the sine of an angle in the second quadrant is positive. So it's no surprise that the sine comes out to be a positive value and, likewise, that the cosine is a negative value (because cosine is negative in the second quadrant).

Overcoming the Differences

By adding angles together, you enlarge your repertoire. You have more *exact* function values — not just the basic function values, but also all the possible sums of these more-common angles. In like fashion, you have even more possibilities for finding the function values of angles when you use subtraction. For example, you can determine the sine of 15 degrees by using 45 degrees and 30 degrees and the appropriate identity.

The subtraction, or difference, identities find the function for the difference between angles α and β:

$$\sin(\alpha - \beta) = \sin\alpha\cos\beta - \cos\alpha\sin\beta$$
$$\cos(\alpha - \beta) = \cos\alpha\cos\beta + \sin\alpha\sin\beta$$
$$\tan(\alpha - \beta) = \frac{\tan\alpha - \tan\beta}{1 + \tan\alpha\tan\beta}$$

Notice how each of the subtraction identities resembles its corresponding angle-sum identity. For the sine rule, the sign between the two products changed from + to –, which seems to make sense. The opposite is true for cosine. The addition rule for cosine has – in it, and the subtraction (or difference) rule has + in it. The tangent rule has both + and – in it; the operation in the numerator mirrors the type of identity.

Only the original three trig functions have truly usable difference identities — the identities for the reciprocal functions are pretty darned complicated. If you want the difference of a reciprocal function, your best bet is to use the corresponding basic identity and find the reciprocal of the numerical answer after you're all finished.

To see one of the subtraction identities in action, check out the following example, which shows how you can find the sine of 15 degrees.

1. **Determine two angles with a difference of 15 degrees.**

   To keep things simple, use 45 and 30.

2. **Substitute the angles into the identity for the sine of a difference.**

   $$\sin(\alpha - \beta) = \sin\alpha\cos\beta - \cos\alpha\sin\beta$$
   $$\sin 15° = \sin(45° - 30°) = \sin 45°\cos 30° - \cos 45°\sin 30°$$

3. **Replace the terms with the function values and simplify the answer.**

   $$\sin(45° - 30°) = \frac{\sqrt{2}}{2}\cdot\frac{1}{2} - \frac{\sqrt{2}}{2}\cdot\frac{\sqrt{3}}{2}$$
   $$= \frac{\sqrt{2}}{4} - \frac{\sqrt{6}}{4} = \frac{\sqrt{2} - \sqrt{6}}{4} = \sin 15°$$

Using radians introduces fractions to the picture, such as finding $\tan\frac{\pi}{12}$ by using the identity for the tangent of a difference.

1. **Determine which angles you need to get the difference.**

   The two angles are $\frac{\pi}{3}$ and $\frac{\pi}{4}$, giving you $\frac{\pi}{3} - \frac{\pi}{4} = \frac{4\pi}{12} - \frac{3\pi}{12} = \frac{\pi}{12}$.

2. **Substitute the angles into the identity for the tangent of a difference.**

   $$\tan(\alpha - \beta) = \frac{\tan\alpha - \tan\beta}{1 + \tan\alpha\tan\beta}$$
   $$\tan\frac{\pi}{12} = \tan\left(\frac{\pi}{3} - \frac{\pi}{4}\right) = \frac{\tan\frac{\pi}{3} - \tan\frac{\pi}{4}}{1 + \tan\frac{\pi}{3}\tan\frac{\pi}{4}}$$

3. **Replace the terms with the function values and simplify the answer.**

   $$\tan\left(\frac{\pi}{3} - \frac{\pi}{4}\right) = \frac{\sqrt{3} - 1}{1 + \sqrt{3}\cdot 1} = \frac{\sqrt{3} - 1}{1 + \sqrt{3}} = \tan\frac{\pi}{12}$$

   The result is rather messy. You can simplify it even more by multiplying the numerator and denominator by the *conjugate* (same terms, different sign) of the denominator and simplifying the result:

   $$\frac{\sqrt{3} - 1}{1 + \sqrt{3}}\cdot\frac{1 - \sqrt{3}}{1 - \sqrt{3}} = \frac{\sqrt{3} - 3 - 1 + \sqrt{3}}{1 - 3} = \frac{2\sqrt{3} - 4}{-2} = -\sqrt{3} + 2$$

In Chapter 10, I explain the opposite-angle identities. This next example uses the identity for the cosine of a difference along with the angle measuring 0 degrees to determine an opposite-angle identity. You may like this explanation better than those in Chapter 10, and it just goes to show you how versatile and user-friendly trig identities are — and how they all get along so well together.

In this example, find $\cos\left(-\frac{\pi}{3}\right)$ by using the identity for the difference between angles.

1. **Determine which angles you need to get the difference.**

   Using 0 and $\frac{\pi}{3}$ and subtracting with the 0 first gives a negative result: $0 - \frac{\pi}{3} = -\frac{\pi}{3}$.

2. **Substitute the angles into the identity for the cosine of a difference.**

   $$\cos(\alpha - \beta) = \cos\alpha\cos\beta + \sin\alpha\sin\beta$$
   $$\cos\left(0 - \frac{\pi}{3}\right) = \cos 0\cos\frac{\pi}{3} + \sin 0\sin\frac{\pi}{3}$$

3. **Replace the angles with the function values and simplify the answer.**

   $$\cos\left(-\frac{\pi}{3}\right) = \cos 0\cos\frac{\pi}{3} + \sin 0\sin\frac{\pi}{3}$$
   $$= 1\cdot\frac{1}{2} + 0\cdot\frac{\sqrt{3}}{2} = \frac{1}{2} = \cos\left(-\frac{\pi}{3}\right)$$

   This answer is exactly what you get if you use the opposite-angle identity for cosine: $\cos(-\theta) = \cos\theta$ or $\cos\left(-\frac{\pi}{3}\right) = \cos\left(\frac{\pi}{3}\right) = \frac{1}{2}$.

In the following example, two negatives make a positive. The angles are positive, but their function values are negative. The two angles in question are α, which is in the fourth quadrant, and β, which is in the third quadrant. The known function values are $\sin\alpha = -\frac{4}{5}$ and $\cos\beta = -\frac{5}{13}$. Find $\cos(\alpha - \beta)$.

1. **Find the necessary function values to calculate the difference.**

   The cosine of the difference of two angles uses both the sine and cosine of each angle involved. You already know the sine of α and cosine of β, so you must determine the cosine of α and the sine of β. Using the Pythagorean identity, you can solve for the missing values:

   - First, use the value for $\sin\alpha$ to solve for $\cos\alpha$:

     $$\sin^2\alpha + \cos^2\alpha = 1$$
     $$\left(-\frac{4}{5}\right)^2 + \cos^2\alpha = 1$$
     $$\cos^2\alpha = 1 - \frac{16}{25} = \frac{9}{25}$$
     $$\cos\alpha = \pm\sqrt{\frac{9}{25}} = \pm\frac{3}{5}$$

   Because angle α is in the fourth quadrant, you know that the cosine of α is positive: $\cos\alpha = \frac{3}{5}$.

• Now use the value for $\cos \beta$ to find $\sin \beta$:

$$\sin^2 \beta + \cos^2 \beta = 1$$

$$\sin^2 \beta + \left(-\frac{5}{13}\right)^2 = 1$$

$$\sin^2 \beta = 1 - \frac{25}{169} = \frac{144}{169}$$

$$\sin \beta = \pm\sqrt{\frac{144}{169}} = \pm\frac{12}{13}$$

Angle β is in the third quadrant, where sine is negative, so $\sin \beta = -\frac{12}{13}$.

2. **Insert the function values into the identity for the cosine of a difference.**

$$\cos(\alpha - \beta) = \cos \alpha \cos \beta + \sin \alpha \sin \beta$$

$$= \left(\frac{3}{5}\right)\left(-\frac{3}{15}\right) + \left(-\frac{4}{5}\right)\left(-\frac{12}{13}\right)$$

3. **Simplify the identity and solve for the answer.**

$$\cos(\alpha - \beta) = \left(-\frac{15}{65}\right) + \left(\frac{48}{65}\right)$$

$$= \frac{33}{65}$$

In the preceding example, the first angle is in the fourth quadrant, so its measure is between 270 and 360 degrees. The other angle, which is in the third quadrant, is between 180 and 270 degrees. The difference between them could be anywhere between 0 and 180 degrees, meaning that the new angle is either in the first or second quadrant. The answer for the cosine of the difference came out positive. Chapter 8 tells you that the cosine is positive in the first quadrant and negative in the second, so the difference between the two angles must be somewhere between 0 and 90 degrees, which means that the new angle is in the first quadrant.

The birthday puzzle

Ever wonder what the likelihood is that two people in a room have the same birthday, assuming that no twins are in the room?

Look around the room the next time you're at a small gathering. How many people would you guess would have to be in the room to have a better than even chance that two people will have the same birthday? Would you guess 364? 182? 91? Not even close. Using probability laws, you have a better than 50 percent chance (over half the time there *will* be a match) if you have only 24 people in a room. Each additional person makes it even more likely that some two people will share their special day.

Doubling Your Money

Identities for angles that are twice as large as one of the common angles are used a lot in calculus and various math, physics, and science disciplines. These identities allow you to deal with a larger angle in the terms of a smaller and more-manageable one. A *double-angle* function is written, for example, as sin 2θ, cos 2α, or tan $2x$, where 2θ, 2α, and $2x$ are the angle measures and the assumption is that you mean sin (2θ), cos (2α), or tan $(2x)$. In this section, I show you how the double-angle formulas for sine and cosine came to be. I don't go off on a tangent here, but all you need to know is that because tangent is equal to the ratio of sine and cosine, its identity comes from their double-angle identities.

The double-angle identities find the function for twice the angle θ. Note that the cosine function has three different versions of its double-angle identity.

$$\sin 2\theta = 2\sin\theta\cos\theta$$
$$\cos 2\theta = \cos^2\theta - \sin^2\theta$$
$$= 2\cos^2\theta - 1$$
$$= 1 - 2\sin^2\theta$$
$$\tan 2\theta = \frac{2\tan\theta}{1 - \tan^2\theta}$$

One plus one equals two sines

To show you where the double-angle formula for sine comes from, I use the identity for the sine of a sum, $\sin(\alpha + \beta) = \sin\alpha\cos\beta + \cos\alpha\sin\beta$. If $\alpha = \beta$, then I can replace β with α in the formula, giving me

$$\sin(\alpha + \alpha) = \sin\alpha\cos\alpha + \cos\alpha\sin\alpha = \sin\alpha\cos\alpha + \sin\alpha\cos\alpha$$
$$\sin(2\alpha) = 2\sin\alpha\cos\alpha$$

For example, you can use this double-angle identity to find the function value for the sine of 180 degrees.

1. **Determine which angle is half of 180 degrees.**

   Twice 90 is 180, so the choice is 90 degrees.

2. **Substitute the measure into the double-angle identity for sine.**

   $\sin 180° = \sin 2(90°) = 2\sin 90°\cos 90°$

3. **Replace the angles with the function values and simplify the answer.**

   $\sin 180° = 2(1)(0) = 0$

Unlikely mathematician

Napoleon Bonaparte is best known for his triumphs and trials in French history. But did you know that he was a closet mathematician? He even has a rule named for him. Napoleon is credited with discovering that when you construct equilateral triangles on the sides of any other triangle and then join the centers of those triangles with segments, those segments form another equilateral triangle. Here are some pictures illustrating Napoleon's theorem:

Three's a crowd

Finding the cosine of twice an angle is easier than finding the other function values, because you have three choices to use. You make your choice depending on what information is available and what looks easiest. To show you where the first of the double-angle identities for cosine comes from, I use the angle-sum identity for cosine. Because the two angles are equal, you can replace β with α, so $\cos(\alpha + \beta) = \cos\alpha\cos\beta - \sin\alpha\sin\beta$ becomes

$$\cos(\alpha + \alpha) = \cos\alpha\cos\alpha - \sin\alpha\sin\alpha = (\cos\alpha)^2 - (\sin\alpha)^2$$
$$\cos(2\alpha) = \cos^2\alpha - \sin^2\alpha$$

To get the second version, use the first Pythagorean identity, $\sin^2 \alpha + \cos^2 \alpha = 1$. Solving for $\sin^2 \alpha$, you get $\sin^2 \alpha = 1 - \cos^2 \alpha$. Putting this result back into the double-angle identity for cosine and simplifying, you get

$$\cos(2\alpha) = \cos^2 \alpha - (1 - \cos^2 \alpha)$$
$$= \cos^2 \alpha - 1 + \cos^2 \alpha$$
$$= 2\cos^2 \alpha - 1$$

To find the last version of the double-angle identity for cosine, solve the first Pythagorean identity for $\cos^2 \alpha$, which gives you $\cos^2 \alpha = 1 - \sin^2 \alpha$. Then substitute this result into the first angle-sum identity for cosine:

$$\cos(2\alpha) = (1 - \sin^2 \alpha) - (\sin^2 \alpha)$$
$$= 1 - \sin^2 \alpha - \sin^2 \alpha$$
$$= 1 - 2\sin^2 \alpha$$

The biggest advantage to having three different identities for the cosine of a double angle is that you can solve for the cosine with just one other function value. The sum and difference identities for sine and cosine, on the other hand, as well as the double-angle identity for sine, all involve both the sine and cosine of the angles.

Here's an example showing off that advantage. Find $\cos 2\alpha$; the angle α is in the fourth quadrant, and $\sin \alpha = -0.45$.

1. **Choose the appropriate double-angle identity.**

   Because you know the value of the sine, use $\cos 2\alpha = 1 - 2\sin^2 \alpha$.

2. **Insert the given value in the formula and simplify.**

   $$\cos 2\alpha = 1 - 2(-0.45)^2$$
   $$= 1 - 2(0.2025) = 1 - 0.4050 = 0.5950$$

The resulting cosine is positive. The cosine is positive in the first and fourth quadrants, so how do you know which of those two quadrants the terminal side of this double angle lies in? Go back to the beginning of the problem — you know that the original angle is in the fourth quadrant. An angle in QIV measures between 270 degrees and 360 degrees. If you double those numbers (because you're working with a double angle), you get 540 degrees and 720 degrees. The angles between those two values lie in the third and fourth quadrants. The cosine is positive in the fourth quadrant, so this double angle lies in the fourth quadrant.

Halving Fun Yet?

The trig identities come in sums, differences, multiples, and halves. With these identities, you can get the value of a sine for a 15-degree angle by taking half of 30 degrees. You can also get the value of the tangent of a 22½-degree angle by using half of 45 degrees. These identities just create more and more ways to establish an exact value for many of the more commonly used trig functions.

The half-angle identities find the function value for half the measure of angle θ:

$$\sin\frac{\theta}{2} = \pm\sqrt{\frac{1-\cos\theta}{2}}$$

$$\cos\frac{\theta}{2} = \pm\sqrt{\frac{1+\cos\theta}{2}}$$

$$\tan\frac{\theta}{2} = \frac{\sin\theta}{1+\cos\theta} = \frac{1-\cos\theta}{\sin\theta}$$

The half-angle identities are a result of taking the double-angle identities and scrunching them around. A more-technical term for *scrunching* is to solve for the single angle in a double-angle identity. Here's how the half-angle identity for sine came to be:

1. **Write the double-angle identity for cosine that has just a sine in it.**

   $\cos 2\theta = 1 - 2\sin^2\theta$

   Using the double-angle identity for cosine works better than the double-angle identity for sine, because the sine formula has both functions on the right side of the equation, and you can't easily get rid of one or the other.

2. **Solve for sin θ. First, get the $\sin^2\theta$ term by itself on the left.**

   $\cos 2\theta = 1 - 2\sin^2\theta$

   $2\sin^2\theta = 1 - \cos 2\theta$

3. **Divide each side by 2 and then take the square root of each side.**

   $\sin^2\theta = \dfrac{1-\cos 2\theta}{2}$

   $\sqrt{\sin^2\theta} = \pm\sqrt{\dfrac{1-\cos 2\theta}{2}}$

   $\sin\theta = \pm\sqrt{\dfrac{1-\cos 2\theta}{2}}$

4. **Replace 2θ with α and θ with $\frac{\alpha}{2}$.**

   $\sin\dfrac{\alpha}{2} = \pm\sqrt{\dfrac{1-\cos\alpha}{2}}$

   By switching the letters, you can see the relationship between the two angles, that one is half as big as the other, more easily.

You can find the half-angle formula for cosine by using the appropriate choice of cosine's double-angle identity and following similar steps to the preceding ones. With tangent, you use both the sine and cosine identities. But first, what's this business of + or − in the sine and cosine half-angle identities?

Explaining the ±

Trig identities are numerous. I list most of them somewhere in this chapter and in Chapter 10. You may be wondering, however, why these half-angle identities are in a league of their own — some of them have ± in front of them. Continue on, dear reader, for the answer to that nagging question.

What's unique about the half-angle identities for sine and cosine is the fact that the sign attached depends on what quadrant the angle that you're cutting in half is in or how big that angle is. If you want to know the sine of half a 30-degree angle, you look at what quadrant that half is in. Both 30 degrees and its half, 15 degrees, are in the first quadrant, so their sines are both positive. Not so with the sines of 300 degrees and its half, 150 degrees, though. The sine is negative in the fourth quadrant, where the 300-degree angle is located (in standard position), but the sine of 150 degrees is positive. When you apply the half-angle formulas, you have to consider which quadrant each angle is in and apply the appropriate signs.

Half a tangent is double the fun

The half-angle identity for tangent has two versions. Rather than being a nuisance, having more than one option is really rather nice, because you can choose the version that works best for your situation. The half-angle formulas for the tangent involve both sine and cosine, but those functions switch places in the numerator and denominator of the fraction. Sometimes the sine of a function doesn't have a radical in its exact value when the cosine does (or vice versa). Depending on the sine and cosine values, you choose the version of the half-angle tangent identity that'll be easiest to work with after you input the values. The math is easier when you don't have to worry about those radicals in the denominator.

First, where do these half-angle tangent identities come from?

1. **Use the ratio identity for tangent and fill in the half-angle identities for sine and cosine.**

$$\tan \frac{\theta}{2} = \frac{\sin \frac{\theta}{2}}{\cos \frac{\theta}{2}} = \frac{\sqrt{\frac{1 - \cos \theta}{2}}}{\sqrt{\frac{1 + \cos \theta}{2}}}$$

You can leave off the $\pm$ sign because you won't have to choose which sign to use with the tangent identity — the square-root sign squares out of the equation.

2. **Put the numerator and denominator under the same radical and then simplify the complex fraction.**

$$= \sqrt{\frac{\dfrac{1 - \cos\theta}{2}}{\dfrac{1 + \cos\theta}{2}}} = \sqrt{\frac{1 - \cos\theta}{2} \cdot \frac{2}{1 + \cos\theta}} = \sqrt{\frac{1 - \cos\theta}{1 + \cos\theta}}$$

3. **Multiply the numerator and denominator by the** *conjugate* **(same terms, different sign) of the denominator.**

$$= \sqrt{\frac{1 - \cos\theta}{1 + \cos\theta} \cdot \frac{1 - \cos\theta}{1 - \cos\theta}} = \sqrt{\frac{(1 - \cos\theta)^2}{1 - \cos^2\theta}}$$

4. **Replace the denominator by using the Pythagorean identity and then simplify by putting the radical over the numerator and denominator, individually.**

$$\sqrt{\frac{(1 - \cos\theta)^2}{1 - \cos^2\theta}} = \sqrt{\frac{(1 - \cos\theta)^2}{\sin^2\theta}}$$

$$= \frac{\sqrt{(1 - \cos\theta)^2}}{\sqrt{\sin^2\theta}} = \frac{1 - \cos\theta}{\sin\theta}$$

To find the other form of this half-angle tangent identity, change the preceding Step 3 by multiplying the numerator and denominator of the fraction by the conjugate of the *numerator* instead of the denominator.

Using half-angle identities

By adding, subtracting, or doubling angle measures, you can find lots of exact values of trig functions. This section provides some examples of the types of angles and their functions that you can find with the half-angle identities.

Pi gone wrong

In 1853, William Shanks published his calculation of the decimal value of pi to 707 decimal places, which he computed all by hand. Not until 1945 did someone discover that the last 180 digits of this computation were wrong. But an even greater error occurred in 1897. That year, the General Assembly of the State of Indiana enacted Bill Number 246, stating that pi was legally equal to 4.

Even though you can use a difference identity to find the sine of 15 degrees, you can also use the half-angle identity.

1. **Determine which angle is double the angle you're working with.**

   Half of 30 is 15, so the choice is 30 degrees. Stick to the more-common angles — the ones that have exact values (see Chapter 7) or are multiples of 30 and 45.

2. **Substitute that angle into the half-angle identity for sine.**

   $$\sin\frac{\theta}{2} = \pm\sqrt{\frac{1-\cos\theta}{2}}$$

   $$\sin 15° = \sin\frac{30°}{2} = +\sqrt{\frac{1-\cos 30°}{2}}$$

   Because the sine of 15 degrees is a positive value, the sign in front of the radical becomes +.

3. **Fill in the function values and simplify the answer.**

   $$\sin 15° = \sqrt{\frac{1-\cos 30°}{2}} = \sqrt{\frac{1-\frac{\sqrt{3}}{2}}{2}} = \sqrt{\frac{\frac{2}{2}-\frac{\sqrt{3}}{2}}{2}} = \sqrt{\frac{2-\sqrt{3}}{4}}$$

   The result isn't a particularly pretty value, although beauty is in the eye of the beholder. Some would consider this answer to be wonderful, because it's the exact value and not a decimal approximation.

Now try using the half-angle identity with radicals. Find $\tan\frac{\pi}{8}$.

1. **Determine which angle is double the angle you're working with.**

   The angle $\frac{\pi}{4}$ is double $\frac{\pi}{8}$.

2. **Substitute the angle measure into one of the half-angle tangent identities.**

   $$\tan\frac{\theta}{2} = \frac{1-\cos\theta}{\sin\theta}$$

   $$\tan\frac{\pi}{8} = \tan\frac{\frac{\pi}{4}}{2} = \frac{1-\cos\frac{\pi}{4}}{\sin\frac{\pi}{4}}$$

3. **Fill in the function values and simplify the answer.**

   $$\tan\frac{\pi}{8} = \frac{1-\frac{\sqrt{2}}{2}}{\frac{\sqrt{2}}{2}} = \frac{\frac{2}{2}-\frac{\sqrt{2}}{2}}{\frac{\sqrt{2}}{2}} = \frac{2-\sqrt{2}}{2} \cdot \frac{2}{\sqrt{2}} = \frac{2-\sqrt{2}}{\sqrt{2}}$$

4. **To get the radical out of the denominator, rationalize it by multiplying both parts of the fraction by the conjugate of the denominator.**

$$\tan \frac{\pi}{8} = \frac{2 - \sqrt{2}}{\sqrt{2}} \cdot \frac{\sqrt{2}}{\sqrt{2}} = \frac{2\sqrt{2} - 2}{2} = \frac{2(\sqrt{2} - 1)}{2} = \sqrt{2} - 1$$

The other identity for the tangent of a half angle gives you exactly the same answer. That form isn't any easier, though, because both the sine and cosine of this angle have a radical in them. If the problem involved an angle of 60 degrees, though, the story would be different. The sine of 60 degrees is $\frac{\sqrt{3}}{2}$, and the cosine is $\frac{1}{2}$, which practically begs you to use the form with the cosine in the denominator so you don't have to mess with a radical in the denominator. Both identities work — the one you use is just a matter of personal preference.

Chapter 12

Proving Identities

*O*ne major aspect that people remember about trigonometry, if they studied it in school, is the time they spent proving identities — making one side of an equation match the other. Many people love these little puzzles. The challenge, followed by success, can be a real thrill. Others, though, find this task of proving identities to be less than exciting. What you find in this chapter are a game plan and suggestions so those who aren't so fond of solving such puzzles may begin to actually enjoy it. For your reading pleasure, I divide this chapter into the methods that work best to prove the different types of identities.

Why do you need to prove identities? Don't you already know that they're correct if they're called identities? Sure you do, but proving them is still helpful down the road when you're solving complex trig problems, because the process to solving them is all in the preparation. Many of the trig expressions that you use to solve practical problems have rather complicated and nasty-looking terms. By doing substitutions and manipulations with the trig identities, you can make those expressions more usable. All this practice with proving identities prepares you and gives you a heads-up as to what's possible.

Lining Up the Players

Before starting to prove (or solve) identities, you need to look over the different equivalences that you use to solve them.

Here's a quick list for your reference:

Reciprocal Identities

$$\frac{1}{\sin\theta} = \csc\theta \qquad \frac{1}{\cos\theta} = \sec\theta \qquad \frac{1}{\tan\theta} = \cot\theta$$

$$\frac{1}{\cot\theta} = \tan\theta \qquad \frac{1}{\sec\theta} = \cos\theta \qquad \frac{1}{\csc\theta} = \sin\theta$$

Ratio Identities

$$\tan\theta = \frac{\sin\theta}{\cos\theta}$$

$$\cot\theta = \frac{\cos\theta}{\sin\theta}$$

Opposite-Angle Identities

$$\sin(-\theta) = -\sin\theta$$
$$\cos(-\theta) = \cos\theta$$
$$\tan(-\theta) = -\tan\theta$$

Pythagorean Identities

$$\sin^2\theta + \cos^2\theta = 1; \ \sin^2\theta = 1 - \cos^2\theta; \ \cos^2\theta = 1 - \sin^2\theta$$
$$\tan^2\theta + 1 = \sec^2\theta; \ \tan^2\theta = \sec^2\theta - 1$$
$$1 + \cot^2\theta = \csc^2\theta; \ \cot^2\theta = \csc^2\theta - 1$$

Sum and Difference Identities

$$\sin(\alpha + \beta) = \sin\alpha\cos\beta + \cos\alpha\sin\beta$$
$$\sin(\alpha - \beta) = \sin\alpha\cos\beta - \cos\alpha\sin\beta$$

$$\cos(\alpha + \beta) = \cos\alpha\cos\beta - \sin\alpha\sin\beta$$
$$\cos(\alpha - \beta) = \cos\alpha\cos\beta + \sin\alpha\sin\beta$$

$$\tan(\alpha + \beta) = \frac{\tan\alpha + \tan\beta}{1 - \tan\alpha\tan\beta}$$
$$\tan(\alpha - \beta) = \frac{\tan\alpha - \tan\beta}{1 + \tan\alpha\tan\beta}$$

Double-Angle Identities

$$\sin 2\theta = 2\sin\theta\cos\theta$$

$$\cos 2\theta = \cos^2\theta - \sin^2\theta = 2\cos^2\theta - 1 = 1 - 2\sin^2\theta$$

$$\tan 2\theta = \frac{2\tan\theta}{1 - \tan^2\theta}$$

Half-Angle Identities

$$\sin\frac{\theta}{2} = \pm\sqrt{\frac{1 - \cos\theta}{2}}$$

$$\cos\frac{\theta}{2} = \pm\sqrt{\frac{1 + \cos\theta}{2}}$$

$$\tan\frac{\theta}{2} = \frac{\sin\theta}{1 + \cos\theta} = \frac{1 - \cos\theta}{\sin\theta}$$

Picking Sides

When you prove identities, you usually work on only one side of the equation or the other — not both at the same time — and for good reason. In other math areas, such as solving antiderivatives in calculus, you need to change from one trig expression to another so you can do the problem; such a situation doesn't have any sides, so you need to work on just the one term or expression. So solving trig identities on one side is good practice. The good news is that in these problems, you usually get to pick which side.

Here are the guidelines for choosing sides. You may take your pick based on any one of them.

✔ Choose the side with the greater number of terms.

✔ Choose the side with factors that you need to multiply together.

✔ Choose the side with terms other than sine or cosine.

✔ Choose the side with fractions that you need to add together.

Of course, these guidelines aren't all the possibilities, but they give you a good start. Enter the puzzle part of identities. You get to look for clues in each identity to help you decide which side to work on, as in the following examples.

Prove the identity $\sec x - \sin x \tan x = \cos x$.

1. **Choose the side to work on.**

   The left side has more terms, and two of the functions aren't sine or cosine, so you use two of the guidelines in making this choice.

2. **Use the reciprocal identity to replace sec x and the ratio identity to replace tan x.**

$$\frac{1}{\cos x} - \sin x \cdot \frac{\sin x}{\cos x} = \cos x$$

3. **Multiply the two factors together in the second term. Then combine the two fractions, because they have a common denominator.**

$$\frac{1}{\cos x} - \frac{\sin x}{1} \cdot \frac{\sin x}{\cos x} = \cos x$$

$$\frac{1}{\cos x} - \frac{\sin^2 x}{\cos x} = \cos x$$

$$\frac{1 - \sin^2 x}{\cos x} = \cos x$$

4. **Replace the numerator by using the Pythagorean identity $\cos^2 x = 1 - \sin^2 x$.**

$$\frac{\cos^2 x}{\cos x} = \cos x$$

5. **Reduce the fraction on the left.**

$$\frac{\cos^{\not{2}} x}{\not{\cos x}} = \cos x$$

$$\cos x = \cos x$$

Next, prove the identity $\dfrac{1 - \cos x}{\sin x} + \dfrac{\sin x}{1 - \cos x} = 2 \csc x$.

1. **Choose the side to work on.**

   The left side has fractions that you need to add together.

2. **Find the common denominator.**

   The fractions have two different denominators, so multiply each by a fraction that equals 1 — the fraction with the other fraction's denominator in both the numerator and denominator.

$$\frac{1 - \cos x}{\sin x} \cdot \frac{1 - \cos x}{1 - \cos x} + \frac{\sin x}{1 - \cos x} \cdot \frac{\sin x}{\sin x} = 2 \csc x$$

3. **Simplify the two fractions. Then add them together, because they have the same denominators.**

$$\frac{(1 - \cos x)^2}{\sin x \,(1 - \cos x)} + \frac{\sin^2 x}{\sin x \,(1 - \cos x)} = 2 \csc x$$

$$\frac{(1 - \cos x)^2 + \sin^2 x}{\sin x \,(1 - \cos x)} = 2 \csc x$$

4. **Multiply out the squared part in the far-left term of the numerator.**

$$\frac{1 - 2\cos x + \cos^2 x + \sin^2 x}{\sin x\,(1 - \cos x)} = 2\csc x$$

5. **Replace the last two terms in the numerator with 1, using the Pythagorean identity $\sin^2 x + \cos^2 x = 1$.**

$$\frac{1 - 2\cos x + 1}{\sin x\,(1 - \cos x)} = 2\csc x$$

6. **Combine the two 1s in the numerator, and then factor out the 2 from each of the terms.**

$$\frac{2 - 2\cos x}{\sin x\,(1 - \cos x)} = 2\csc x$$

$$\frac{2\,(1 - \cos x)}{\sin x\,(1 - \cos x)} = 2\csc x$$

7. **Cancel out the common factor in the numerator and denominator.**

$$\frac{2\,(1 - \cancel{\cos x})}{\sin x\,(1 - \cancel{\cos x})} = 2\csc x$$

$$\frac{2}{\sin x} = 2\csc x$$

8. **Now just use the reciprocal identity for the sine to finish up.**

$$\frac{1}{\sin x} = \csc x, \text{ so } \frac{2}{\sin x} = 2\csc x$$

The next example uses the multiplying-out guideline and the Pythagorean identity to make for a pretty result.

Prove the identity that $\cos x\,(\sec x - \cos x) = \sin^2 x$.

1. **Decide which side you'll work on.**

   The left side just begs to be multiplied out by distributing $\cos x$ over the two terms in the parentheses.

   $$\cos x \sec x - \cos^2 x = \sin^2 x$$

2. **Cosine and secant are reciprocals, so their product is 1. Replace the term "$\cos x \sec x$" with 1.**

   $$1 - \cos^2 x = \sin^2 x$$

3. **Now just use the Pythagorean identity to replace the terms on the left.**

   $$\sin^2 x = \sin^2 x$$

Working on Both Sides

As much fun as it is to work on just one side of an identity, sometimes working on both sides at the same time is advantageous and permissible. Working on both sides of an identity is often necessary when you don't have a clear-cut way to get from one side to the other. I've even had to resort to working on both sides when it *wasn't* permissible; working backwards from one side to the result on the other side can give some valuable clues on how to solve the thing.

With a trig identity, working on both sides isn't really the same as working on both sides of an algebraic equation. In algebra, you can multiply each side by the same number, square both sides, add or subtract the same thing to each side, and so on. When you solve trig identities and equations (see Chapter 14), you can use all those algebra rules *plus* you can do substitutions with the various trig identities when you need them. You can even insert a different identity on each side — the one big advantage of working on both sides of a trig identity.

This first example is rather basic, but it gets the idea across. Solve the identity $\frac{\sin\theta}{\csc\theta} + \frac{\cos\theta}{\sec\theta} = \tan\theta\cot\theta$ by working on both sides.

1. **Change the two denominators of the fractions by using their reciprocal identities. Also change the cotangent on the right by using its reciprocal.**

   $$\frac{\sin\theta}{\frac{1}{\sin\theta}} + \frac{\cos\theta}{\frac{1}{\cos\theta}} = \tan\theta \cdot \frac{1}{\tan\theta}$$

2. **Simplify the two fractions on the left by flipping the denominators and multiplying them by their numerators. Then multiply the two factors on the right together.**

   $$\sin\theta \cdot \frac{\sin\theta}{1} + \cos\theta \cdot \frac{\cos\theta}{1} = 1$$
   $$\frac{\sin\theta}{1} \cdot \frac{\sin\theta}{1} + \frac{\cos\theta}{1} \cdot \frac{\cos\theta}{1} = 1$$
   $$\sin^2\theta + \cos^2\theta = 1$$

3. **Replace the sum on the left by using the Pythagorean identity.**

   You end up with $1 = 1$.

In the next example, you change everything to sines and cosines. Prove the identity $\frac{\csc x}{\cos x} = \cot x + \tan x$.

1. **Change the functions to their equivalences by using the reciprocal and ratio identities.**

$$\frac{1}{\frac{\sin x}{\cos x}} = \frac{\cos x}{\sin x} + \frac{\sin x}{\cos x}$$

2. **On the left, flip the denominator and multiply it by the numerator. On the right, multiply each fraction by a fraction equal to 1 (by using the other fraction's denominator) to get common denominators for all the fractions.**

$$\frac{1}{\sin x} \cdot \frac{1}{\cos x} = \frac{\cos x}{\sin x} \cdot \frac{\cos x}{\cos x} + \frac{\sin x}{\cos x} \cdot \frac{\sin x}{\sin x}$$

3. **Simplify the multiplied fractions. Add the two fractions on the right together.**

$$\frac{1}{\sin x \cos x} = \frac{\cos^2 x}{\sin x \cos x} + \frac{\sin^2 x}{\sin x \cos x}$$
$$= \frac{\cos^2 x + \sin^2 x}{\sin x \cos x}$$

4. **Replace the numerator on the right with the value from the Pythagorean identity.**

$$\frac{1}{\sin x \cos x} = \frac{1}{\sin x \cos x}$$

This last example requires a little creativity to get the job done. But working on both sides still works best when solving $\frac{1 + \cot \alpha}{\cot \alpha} = \tan \alpha + \csc^2 \alpha - \cot^2 \alpha$.

1. **Split up the fraction on the left by writing each term in the numerator over the denominator.**

$$\frac{1}{\cot \alpha} + \frac{\cot \alpha}{\cot \alpha} = \tan \alpha + \csc^2 \alpha - \cot^2 \alpha$$

2. **Reduce the second fraction to 1.**

$$\frac{1}{\cot \alpha} + 1 = \tan \alpha + \csc^2 \alpha - \cot^2 \alpha$$

3. **Now replace the $\csc^2 \alpha$ on the right with its equivalent by using the Pythagorean identity.**

$$\frac{1}{\cot \alpha} + 1 = \tan \alpha + \left(1 + \cot^2 \alpha\right) - \cot^2 \alpha$$

4. **Simplify the terms on the right — two are opposites of one another.**

$$\frac{1}{\cot \alpha} + 1 = \tan \alpha + 1 + \cancel{\cot^2 \alpha} - \cancel{\cot^2 \alpha}$$
$$= \tan \alpha + 1$$

5. **Replace the fraction on the left by using the reciprocal identity.**

$$\tan \alpha + 1 = \tan \alpha + 1$$

The music of the spheres

Pythagoras is best known for his theorem, which defines the relationships among the lengths of a right triangle's sides, but his second most well-known contribution to humanity is his discovery of the mathematical basis of the musical scale. He found that a connection exists between musical harmony — the stuff that sounds good — and whole numbers. If you pluck a taut string, listen to the note, and then pluck a string twice as long and equally taut, you hear a note one octave below the first note. You can also go down the scale by increasing the length of the taut string in smaller increments.

Pythagoras believed that whole-number relationships represent all harmony, all beauty, and all nature. He extended this theory to the orbits of the planets and believed that as the planets move through space, they must give off a heavenly whole-number harmony. Hence the term *the music of the spheres.*

Going Back to Square One

With some identities, which side you should work on or what you should do with either or both sides isn't clear. And in some instances, you're faced with such a conglomeration of functions that figuring out what's going on is darn near impossible. Other times, the different terms have different powers of the same function. In such cases, simplifying matters by either changing everything to sines and cosines or by factoring out some function may be your best bet.

Changing to sines and cosines

In this first example, you can use either reciprocal or ratio identities, depending on which side you're going to work on, to change everything to sines and cosines. Here's how I'd solve the identity $\tan\theta + \cot\theta = \csc\theta\sec\theta$:

1. **Replace the two terms on the left by using ratio identities.**

$$\frac{\sin\theta}{\cos\theta} + \frac{\cos\theta}{\sin\theta} = \csc\theta\sec\theta$$

2. **To get a common denominator, multiply both terms on the left by fractions equal to 1 (by using the other term's denominator).**

$$\frac{\sin\theta}{\cos\theta} \cdot \frac{\sin\theta}{\sin\theta} + \frac{\cos\theta}{\sin\theta} \cdot \frac{\cos\theta}{\cos\theta} = \csc\theta\sec\theta$$

3. **Simplify the fractions and then add them together, because they have a common denominator.**

$$\frac{\sin^2\theta}{\cos\theta\sin\theta} + \frac{\cos^2\theta}{\sin\theta\cos\theta} = \csc\theta\sec\theta$$

$$\frac{\sin^2\theta + \cos^2\theta}{\cos\theta\sin\theta} = \csc\theta\sec\theta$$

4. **Replace the numerator on the left with its equivalent by using the Pythagorean identity.**

$$\frac{1}{\cos\theta\sin\theta} = \csc\theta\sec\theta$$

5. **Now use the reciprocal identities on the denominator and then flip each fraction and multiply.**

$$\frac{1}{\frac{1}{\sec\theta}\cdot\frac{1}{\csc\theta}} = \csc\theta\sec\theta$$

$$\frac{\sec\theta}{1}\cdot\frac{\csc\theta}{1} = \csc\theta\sec\theta$$

In the next example, only two terms aren't already written as sines, so switching those two to sines just seems natural when solving $\frac{\sin x + 8\csc x}{\sin x + 4\csc x} = \frac{\sin^2 x + 8}{\sin^2 x + 4}$.

1. **Change the two cosecants by using the reciprocal identity.**

$$\frac{\sin x + 8\left(\frac{1}{\sin x}\right)}{\sin x + 4\left(\frac{1}{\sin x}\right)} = \frac{\sin^2 x + 8}{\sin^2 x + 4}$$

$$\frac{\sin x + \frac{8}{\sin x}}{\sin x + \frac{4}{\sin x}} = \frac{\sin^2 x + 8}{\sin^2 x + 4}$$

2. **Multiply each term in the numerator and denominator by sin x. This action amounts to multiplying by sine over sine, or by 1.**

$$\frac{\sin x\left(\sin x + \frac{8}{\sin x}\right)}{\sin x\left(\sin x + \frac{4}{\sin x}\right)} = \frac{\sin^2 x + 8}{\sin^2 x + 4}$$

$$\frac{\sin x \cdot \sin x + \sin x \cdot \frac{8}{\sin x}}{\sin x \cdot \sin x + \sin x \cdot \frac{4}{\sin x}} = \frac{\sin^2 x + 8}{\sin^2 x + 4}$$

3. **Simplify the numerator and denominator.**

$$\frac{\sin^2 x + \cancel{\sin x} \cdot \dfrac{8}{\cancel{\sin x}}}{\sin^2 x + \cancel{\sin x} \cdot \dfrac{4}{\cancel{\sin x}}} = \frac{\sin^2 x + 8}{\sin^2 x + 4}$$

$$\frac{\sin^2 x + 8}{\sin^2 x + 4} = \frac{\sin^2 x + 8}{\sin^2 x + 4}$$

This last example has so many functions and terms involved that figuring out where to start almost seems impossible. Although you have other ways to approach it, I change the fraction on the left to all sines and cosines. If you want to see another way to solve an identity like this one, refer to the "Finding a common denominator" section, later in this chapter.

Solve the identity $\dfrac{1 + \sec x}{\tan x} - \dfrac{\tan x}{\sec x} = \cot x \, (1 + \cos x)$.

1. **On the left side, change the secants by using the reciprocal identity and the tangents by using the ratio identity.**

$$\frac{1 + \dfrac{1}{\cos x}}{\dfrac{\sin x}{\cos x}} - \frac{\dfrac{\sin x}{\cos x}}{\dfrac{1}{\cos x}} = \cot x \, (1 + \cos x)$$

2. **Multiply each term in the numerator and denominator by cos x and simplify all the terms.**

$$\frac{\cos x \left(1 + \dfrac{1}{\cos x}\right)}{\cos x \left(\dfrac{\sin x}{\cos x}\right)} - \frac{\cos x \left(\dfrac{\sin x}{\cos x}\right)}{\cos x \left(\dfrac{1}{\cos x}\right)} = \cot x \, (1 + \cos x)$$

$$\frac{\cos x \cdot 1 + \cos x \cdot \dfrac{1}{\cos x}}{\cos x \cdot \dfrac{\sin x}{\cos x}} - \frac{\cos x \cdot \dfrac{\sin x}{\cos x}}{\cos x \cdot \dfrac{1}{\cos x}} = \cot x \, (1 + \cos x)$$

$$\frac{\cos x + \cancel{\cos x} \cdot \dfrac{1}{\cancel{\cos x}}}{\cancel{\cos x} \cdot \dfrac{\sin x}{\cancel{\cos x}}} - \frac{\cancel{\cos x} \cdot \dfrac{\sin x}{\cancel{\cos x}}}{\cancel{\cos x} \cdot \dfrac{1}{\cancel{\cos x}}} = \cot x \, (1 + \cos x)$$

From that mess, you get $\dfrac{\cos x + 1}{\sin x} - \dfrac{\sin x}{1} = \cot x \, (1 + \cos x)$.

3. **Find a common denominator for the two fractions on the left, add the fractions together, and simplify the result.**

$$\frac{\cos x + 1}{\sin x} - \frac{\sin x}{\sin x} \cdot \frac{\sin x}{1} = \cot x \, (1 + \cos x)$$

$$\frac{\cos x + 1}{\sin x} - \frac{\sin^2 x}{\sin x} = \cot x \, (1 + \cos x)$$

$$\frac{\cos x + 1 - \sin^2 x}{\sin x} = \cot x \, (1 + \cos x)$$

4. **Now replace $\sin^2 x$ with its equivalent by using the Pythagorean identity, and simplify.**

$$\frac{\cos x + 1 - (1 - \cos^2 x)}{\sin x} = \cot x \,(1 + \cos x)$$

$$\frac{\cos x + 1 - 1 + \cos^2 x}{\sin x} = \cot x \,(1 + \cos x)$$

$$\frac{\cos x + \cos^2 x}{\sin x} = \cot x \,(1 + \cos x)$$

5. **Factor a $\cos x$ from each term in the numerator.**

$$\frac{\cos x \,(1 + \cos x)}{\sin x} = \cot x \,(1 + \cos x)$$

6. **Finally, split the two factors in the numerator into two fractions that are multiplied by each other. Then replace $\frac{\cos x}{\sin x}$ by using the ratio identity.**

$$\frac{\cos x}{\sin x} \cdot \frac{(1 + \cos x)}{1} = \cot x \,(1 + \cos x)$$

$$\cot x \,(1 + \cos x) = \cot x \,(1 + \cos x)$$

Factoring

The clue you'll get that says you need to factor an identity is when powers of a particular function or repeats of that same function are in all the terms on one side of the identity.

For example, the identity $\sin^4 \theta + 2\sin^2 \theta \cos^2 \theta + \cos^4 \theta = 1$ has three terms on the left that you can factor, because they're the result of squaring a binomial. The pattern you need is the algebraic equation $a^2 + 2ab + b^2 = (a + b)^2$.

1. **Factor the expression on the left as the square of a binomial.**

$$\left(\sin^2 \theta + \cos^2 \theta \right)^2 = 1$$

2. **Now just replace the expression in the parentheses with its equivalent by using the Pythagorean identity.**

$$(1)^2 = 1$$

The preceding example was really simple — as long as you recognized the pattern. It'd be another thing altogether if you went off on some tangent (pardon the pun).

In the next example, the factoring occurs in the numerator of the fraction, where powers of $\sin x$ appear. Solve the identity $\frac{\sin x - \sin^3 x}{\cos x} = \tan x \cos^2 x$.

1. **Factor sin x out of each term in the numerator.**

$$\frac{\sin x \left(1 - \sin^2 x\right)}{\cos x} = \tan x \cos^2 x$$

2. **Replace the expression in the parentheses with its equivalent by using the Pythagorean identity.**

$$\frac{\sin x \left(\cos^2 x\right)}{\cos x} = \tan x \cos^2 x$$

3. **Now split up the fraction into a product of two fractions, carefully arranging the numerator and denominator.**

$$\frac{\sin x}{\cos x} \cdot \frac{\left(\cos^2 x\right)}{1} = \tan x \cos^2 x$$

4. **Replace the first fraction with tan x by using the ratio identity.**

$$\tan x \cos^2 x = \tan x \cos^2 x$$

This last example requires factoring by using the difference between two squares. The pattern here is the algebraic equation $a^2 - b^2 = (a - b)(a + b)$ or $a^4 - b^4 = (a^2 - b^2)(a^2 + b^2)$. Solve the identity $\csc^2 \theta + \cot^2 \theta = \csc^4 \theta - \cot^4 \theta$.

1. **Factor the two terms on the right by using the difference-of-squares equation.**

$$\csc^2 \theta + \cot^2 \theta = \left(\csc^2 \theta - \cot^2 \theta\right)\left(\csc^2 \theta + \cot^2 \theta\right)$$

2. **In the left set of parentheses only, replace the csc$^2 \theta$ with its equivalent in the Pythagorean identity.**

   You want to keep the two terms in the right parentheses.

   $$\csc^2 \theta + \cot^2 \theta = \left(1 + \cot^2 \theta - \cot^2 \theta\right)\left(\csc^2 \theta + \cot^2 \theta\right)$$

3. **Now simplify the expression, getting rid of the two opposites.**

   $$\csc^2 \theta + \cot^2 \theta = \left(1 + \cancel{\cot^2 \theta} - \cancel{\cot^2 \theta}\right)\left(\csc^2 \theta + \cot^2 \theta\right)$$
   $$\csc^2 \theta + \cot^2 \theta = 1 \cdot \left(\csc^2 \theta + \cot^2 \theta\right)$$

Using a little bit of both

The examples in this section involve both changing the terms to sines and cosines as well as factoring. The hardest part is deciding what to do first.

In this first example, your work goes more smoothly if you change everything to sines and cosines first. (Plus you may not recognize right away that the expression on the left is the result of squaring a binomial.) Solve the identity

$$\csc^2 \theta - 2\csc \theta \cot \theta + \cot^2 \theta = \frac{1 - \cos \theta}{1 + \cos \theta}.$$

1. **Change the terms on the left to sines and cosines by using reciprocal and ratio identities and then simplify the fractions.**

$$\frac{1}{\sin^2\theta} - \frac{2}{\sin\theta}\cdot\frac{\cos\theta}{\sin\theta} + \frac{\cos^2\theta}{\sin^2\theta} = \frac{1-\cos\theta}{1+\cos\theta}$$

$$\frac{1}{\sin^2\theta} - \frac{2\cos\theta}{\sin^2\theta} + \frac{\cos^2\theta}{\sin^2\theta} = \frac{1-\cos\theta}{1+\cos\theta}$$

2. **Add the three fractions together, because they have the same denominator.**

$$\frac{1-2\cos\theta+\cos^2\theta}{\sin^2\theta} = \frac{1-\cos\theta}{1+\cos\theta}$$

3. **Replace the denominator of the fraction by using the Pythagorean identity.**

$$\frac{1-2\cos\theta+\cos^2\theta}{1-\cos^2\theta} = \frac{1-\cos\theta}{1+\cos\theta}$$

4. **Factor the numerator as the square of a binomial; factor the denominator as the difference of two squares.**

$$\frac{(1-\cos\theta)^2}{(1-\cos\theta)(1+\cos\theta)} = \frac{1-\cos\theta}{1+\cos\theta}$$

5. **Factor out the common binomial in the numerator and denominator.**

$$\frac{(1-\cos\theta)^{\cancel{2}}}{\cancel{(1-\cos\theta)}(1+\cos\theta)} = \frac{1-\cos\theta}{1+\cos\theta}$$

$$\frac{1-\cos\theta}{1+\cos\theta} = \frac{1-\cos\theta}{1+\cos\theta}$$

In the next example, you see how to first do the factoring and then go to the basics. Solve the identity $\dfrac{\sin\theta}{\cot\theta\cos^2\theta - \cot\theta} - \sec\theta = 0$.

1. **Factor cot θ out of each term in the denominator of the first fraction.**

$$\frac{\sin\theta}{\cot\theta\left(\cos^2\theta - 1\right)} - \sec\theta = 0$$

2. **Replace the value in the parentheses with its equivalent by using the Pythagorean identity.**

$$\frac{\sin\theta}{\cot\theta\left(\sin^2\theta\right)} - \sec\theta = 0$$

3. **Write everything in terms of sine and cosine by using reciprocal and ratio identities.**

$$\frac{\sin\theta}{\frac{\cos\theta}{\sin\theta}\left(\sin^2\theta\right)} - \frac{1}{\cos\theta} = 0$$

 4. **Reduce the fractions in the denominator and simplify.**

$$\frac{\sin\theta}{\frac{\cos\theta}{\sin\theta}\left(\frac{\sin^2\theta}{1}\right)} - \frac{1}{\cos\theta} = 0$$

$$\frac{\sin\theta}{\cos\theta\sin\theta} - \frac{1}{\cos\theta} = 0$$

 5. **Cancel out sin θ from the first fraction and simplify.**

$$\frac{\sin\theta}{\cos\theta\sin\theta} - \frac{1}{\cos\theta} = 0$$

$$\frac{1}{\cos\theta} - \frac{1}{\cos\theta} = 0$$

Fracturing Fractions

The ratio and reciprocal identities involve fractions. The half-angle identities use fractions. You just can't get away from them. Actually, an identity with fractions can work to your advantage. You can work toward getting rid of the fraction and, in the process, solve the problem. Some of the main techniques for working with fractions in identities are to either break them up into separate terms or to go in the other direction and find a common denominator.

Breaking up is hard to do

This section's heading is very misleading. Breaking up fractions isn't hard to do. In fact, when you can, doing so is one of the easiest and most productive ways to solve identities. The trick is to break them up correctly. You can break up a fraction with several terms in the numerator and one in the denominator — but not the other way around.

Correct: $\dfrac{1+a-b}{7} = \dfrac{1}{7} + \dfrac{a}{7} - \dfrac{b}{7}$

Incorrect: $\dfrac{3+a-z}{2+b+c} \neq \dfrac{3}{2} + \dfrac{a}{b} - \dfrac{z}{c}$

Now apply this breaking up of fractions to a trig identity. In this first example, the fraction on the left has just one term in the denominator. Solve the identity $\dfrac{\sin x + \cot x}{\cos x} = \tan x + \csc x$.

 1. **Break up the fraction by writing each term over the denominator.**

$$\frac{\sin x}{\cos x} + \frac{\cot x}{\cos x} = \tan x + \csc x$$

2. **Rewrite cot x by using the ratio identity.**

$$\frac{\sin x}{\cos x} + \frac{\frac{\cos x}{\sin x}}{\frac{\sin x}{\cos x}} = \tan x + \csc x$$

3. **Simplify the complex fraction by flipping the denominator and multiplying it times the numerator. Then reduce the result.**

$$\frac{\sin x}{\cos x} + \frac{\cos x}{\sin x} \cdot \frac{1}{\cos x} = \tan x + \csc x$$

$$\frac{\sin x}{\cos x} + \frac{1}{\sin x} = \tan x + \csc x$$

4. **Replace the first fraction by using the ratio identity for tangent and the second fraction by using the reciprocal identity for cosecant.**

$$\tan x + \csc x = \tan x + \csc x$$

You can also break up fractions that have more than one factor (but only one term) in the denominator by carrying them both along. For example, solve the identity $\frac{\cot x - \cos x}{\cot x \cos x} = \frac{1 - \sin x}{\cos x}$.

1. **Break up the fraction by writing each term in the numerator over the entire denominator.**

$$\frac{\cot x}{\cot x \cos x} - \frac{\cos x}{\cot x \cos x} = \frac{1 - \sin x}{\cos x}$$

2. **Reduce each fraction on the left side.**

$$\frac{\cot x}{\cot x \cos x} - \frac{\cos x}{\cot x \cos x} = \frac{1 - \sin x}{\cos x}$$

$$\frac{1}{\cos x} - \frac{1}{\cot x} = \frac{1 - \sin x}{\cos x}$$

3. **Rewrite cot x in the second denominator by using the ratio identity. Then simplify the complex fraction by flipping the denominator and multiplying.**

$$\frac{1}{\cos x} - \frac{1}{\frac{\cos x}{\sin x}} = \frac{1 - \sin x}{\cos x}$$

$$\frac{1}{\cos x} - 1 \cdot \frac{\sin x}{\cos x} = \frac{1 - \sin x}{\cos x}$$

4. **The two fractions now have the same denominator. Rewrite the left side as all one fraction.**

$$\frac{1 - \sin x}{\cos x} = \frac{1 - \sin x}{\cos x}$$

Finding a common denominator

Fractions are your friends. You may find this to be unbelievable, but the more you work with trig functions, the more you're swayed to my way of thinking. Finding a common denominator to combine fractions often paves the way to solving an identity.

In the identity $\dfrac{1-\sin\theta}{\cos\theta} - \dfrac{\cos\theta}{1+\sin\theta} = 0$, the two denominators on the left have nothing in common, so you multiply each fraction by the other's denominator — or, rather, by that denominator over itself, which equals 1.

1. **Multiply each fraction on the left by an equivalent of 1 to create a common denominator.**

$$\frac{1-\sin\theta}{\cos\theta} \cdot \frac{1+\sin\theta}{1+\sin\theta} - \frac{\cos\theta}{1+\sin\theta} \cdot \frac{\cos\theta}{\cos\theta} = 0$$

2. **Multiply the fractions together and simplify the numerators. Leave the denominator alone.**

$$\frac{(1-\sin\theta)(1+\sin\theta)}{\cos\theta\,(1+\sin\theta)} - \frac{\cos\theta\cos\theta}{\cos\theta\,(1+\sin\theta)} = 0$$

$$\frac{1-\sin^2\theta}{\cos\theta\,(1+\sin\theta)} - \frac{\cos^2\theta}{\cos\theta\,(1+\sin\theta)} = 0$$

3. **Replace the first numerator with its equivalent by using the Pythagorean identity. The fractions are opposites of one another.**

$$\frac{\cos^2\theta}{\cos\theta\,(1+\sin\theta)} - \frac{\cos^2\theta}{\cos\theta\,(1+\sin\theta)} = 0$$

$$0 = 0$$

In the "Changing to sines and cosines" section, I did a problem using that method and mentioned that you had another option — finding a common denominator. You have to decide which way you think is better. You may even be able to find an easier way to do it. Here you go: Prove the identity $\dfrac{1+\sec x}{\tan x} - \dfrac{\tan x}{\sec x} = \cot x\,(1+\cos x)$ by finding a common denominator.

Archimedes

Considered to be one of the creative geniuses of the ancient world, Archimedes was an Alexandrian mathematician who lived from about 287 to 212 B.C. One of his discoveries in which he took the most pride was a method for calculating the volume of a sphere. He found that the volume is two-thirds the volume of the smallest cylinder that the sphere can fit into. He even requested that a diagram with the cylinder and sphere be engraved on his tombstone.

1. **Multiply each fraction on the left by the equivalent of 1, creating a common denominator.**

$$\frac{1 + \sec x}{\tan x} \cdot \frac{\sec x}{\sec x} - \frac{\tan x}{\sec x} \cdot \frac{\tan x}{\tan x} = \cot x (1 + \cos x)$$

2. **Simplify the numerators by multiplying out the fractions.**

$$\frac{\sec x + \sec^2 x}{\tan x \sec x} - \frac{\tan^2 x}{\tan x \sec x} = \cot x (1 + \cos x)$$

3. **Replace $\tan^2 x$ with its equivalent by using the Pythagorean identity. Then combine the two numerators.**

$$\frac{\sec x + \sec^2 x}{\tan x \sec x} - \frac{\sec^2 x - 1}{\tan x \sec x} = \cot x (1 + \cos x)$$

$$\frac{\sec x + \sec^2 x - \sec^2 x + 1}{\tan x \sec x} = \cot x (1 + \cos x)$$

4. **Simplify the numerator. Then rewrite the left side as the product of two fractions.**

$$\frac{\sec x + 1}{\tan x \sec x} = \cot x (1 + \cos x)$$

$$\frac{1}{\tan x} \cdot \frac{\sec x + 1}{\sec x} = \cot x (1 + \cos x)$$

5. **Multiply the fraction on the right by cos x divided by cos x, which is equivalent to 1.**

$$\frac{1}{\tan x} \cdot \left(\frac{\sec x + 1}{\sec x} \cdot \frac{\cos x}{\cos x} \right) = \cot x (1 + \cos x)$$

6. **Multiply out the second fraction, distributing through the numerator.**

$$\frac{1}{\tan x} \cdot \left(\frac{\sec x \cos x + 1 \cdot \cos x}{\sec x \cos x} \right) = \cot x (1 + \cos x)$$

7. **Because cos x and sec x are reciprocals, their product is 1. (Substitute 1 in both the numerator and the denominator.)**

$$\frac{1}{\tan x} \cdot \left(\frac{1 + \cos x}{1} \right) = \cot x (1 + \cos x)$$

8. **Replace the reciprocal of tan x with cot x.**

$$\cot x (1 + \cos x) = \cot x (1 + \cos x)$$

Using Tricks of the Trig Trade

When proving identities, sometimes the best way to handle them just leaps out at you — and sometimes the best way just stays in hiding. Usually, you can solve an identity in more than one way — the best way, the almost-as-good way, the reasonable way, and the absolutely dreadful way. The best way is the quickest and most efficient. But sometimes you have to pull something out of your hat to accomplish the task of solving a particular identity. These little tricks amount to nothing more than some algebraic manipulation such as multiplying a fraction by a conjugate or squaring both sides of the identity.

Multiplying by a conjugate

First, what in the world is a *conjugate?* In mathematics, a conjugate consists of the same two terms as the first expression, separated by the opposite sign. For instance, the conjugate of $x + \sqrt{y}$ is $x - \sqrt{y}$. In trig, especially, multiplying the numerator and denominator of a fraction by a conjugate can create some really nice results.

Multiplying by a conjugate is a quick, easy way of solving the identity $\dfrac{1}{\sec x - \tan x} = \tan x + \sec x$.

1. **Multiply the numerator and denominator of the fraction on the left by the conjugate of the denominator.**

   $$\frac{1}{\sec x - \tan x} \cdot \frac{\sec x + \tan x}{\sec x + \tan x} = \tan x + \sec x$$

2. **The two denominators multiplied together are the difference of two squares.**

   $$\frac{\tan x + \sec x}{\sec^2 x - \tan^2 x} = \tan x + \sec x$$

3. **Replace $\sec^2 x$ in the denominator with its equivalent by using the Pythagorean identity.**

   $$\frac{\tan x + \sec x}{\tan^2 x + 1 - \tan^2 x} = \tan x + \sec x$$

4. **Simplify the denominator by canceling out the two opposites.**

   $$\frac{\tan x + \sec x}{\tan^2 x + 1 - \tan^2 x} = \tan x + \sec x$$

   $$\tan x + \sec x = \tan x + \sec x$$

In the next example, you have to decide which fraction to multiply the conjugate by. I choose the fraction on the right, because I see the conjugate of the numerator on the right in the denominator on the left. Solve the identity $\dfrac{\tan x}{1 + \cos x} = \dfrac{1 - \cos x}{\sin x \cos x}$.

1. **Multiply the numerator and denominator of the fraction on the right by the conjugate of the numerator.**

$$\frac{\tan x}{1+\cos x} = \frac{1-\cos x}{\sin x \cos x} \cdot \frac{1+\cos x}{1+\cos x}$$

2. **Multiply the fractions together, keeping the parentheses in the denominator.**

$$\frac{\tan x}{1+\cos x} = \frac{1-\cos^2 x}{\sin x \cos x\,(1+\cos x)}$$

3. **Substitute the equivalent from the Pythagorean identity in the numerator of the fraction on the right. Then reduce the fraction.**

$$\frac{\tan x}{1+\cos x} = \frac{\sin^2 x}{\sin x \cos x\,(1+\cos x)}$$

$$\frac{\tan x}{1+\cos x} = \frac{\sin^{\not{2}} x}{\not{\sin x} \cos x\,(1+\cos x)}$$

$$\frac{\tan x}{1+\cos x} = \frac{\sin x}{\cos x\,(1+\cos x)}$$

4. **Rewrite the fraction on the right as a product of two fractions, carefully arranging the factors.**

$$\frac{\tan x}{1+\cos x} = \frac{\sin x}{\cos x} \cdot \frac{1}{(1+\cos x)}$$

5. **Replace the first fraction on the right with its ratio-identity equivalent. Rewrite the expression as one fraction.**

$$\frac{\tan x}{1+\cos x} = \tan x \cdot \frac{1}{(1+\cos x)}$$

$$\frac{\tan x}{1+\cos x} = \frac{\tan x}{1+\cos x}$$

The half-angle identity for the tangent function has two different forms. Multiplying by the conjugate is a good method for showing that these two forms are equivalent. In this example, I prove that $\dfrac{\sin\theta}{1+\cos\theta} = \dfrac{1-\cos\theta}{\sin\theta}$.

1. **Multiply the numerator and denominator of the fraction on the left by the conjugate of the denominator.**

$$\frac{\sin\theta}{1+\cos\theta} \cdot \frac{1-\cos\theta}{1-\cos\theta} = \frac{1-\cos\theta}{\sin\theta}$$

2. **Multiply the two denominators together, but leave the numerator in factored form.**

$$\frac{\sin\theta\,(1-\cos\theta)}{1-\cos^2\theta} = \frac{1-\cos\theta}{\sin\theta}$$

3. **Replace the denominator on the left with its equivalent by using the Pythagorean identity.**

$$\frac{\sin\theta\,(1-\cos\theta)}{\sin^2\theta} = \frac{1-\cos\theta}{\sin\theta}$$

4. **Reduce the fraction on the left.**

$$\frac{\cancel{\sin\theta}\,(1-\cos\theta)}{\sin^{\cancel{2}}\theta} = \frac{1-\cos\theta}{\sin\theta}$$

$$\frac{1-\cos\theta}{\sin\theta} = \frac{1-\cos\theta}{\sin\theta}$$

Squaring both sides

One special case of working on both sides of an identity at the same time is to square both sides. Your biggest clue as to when to use this technique is usually when one side or the other has a radical. This method is also good to use when you're solving some types of trig equations. Squaring both sides has benefits twofold: It gets rid of radicals, and it creates terms that can be part of one of the Pythagorean identities. The Pythagorean identities have wonderful substitutions.

This first example has only one radical, and it's on the right side. Solve the identity $\frac{1-\cot x}{\csc x} = \sqrt{1-2\sin x\cos x}$.

1. **Square both sides of the identity.**

   Be sure to expand the squared binomial on the left correctly.

   $$\left(\frac{1-\cot x}{\csc x}\right)^2 = \left(\sqrt{1-2\sin x\cos x}\right)^2$$

   $$\frac{(1-\cot x)^2}{\csc^2 x} = 1-2\sin x\cos x$$

   $$\frac{1-2\cot x+\cot^2 x}{\csc^2 x} = 1-2\sin x\cos x$$

2. **Rearrange the terms in the numerator.**

   $$\frac{1+\cot^2 x-2\cot x}{\csc^2 x} = 1-2\sin x\cos x$$

3. **Replace $1+\cot^2 x$ with its equivalent by using the Pythagorean identity.**

   $$\frac{\csc^2 x-2\cot x}{\csc^2 x} = 1-2\sin x\cos x$$

4. **Split up the fraction by writing each term in the numerator over the denominator.**

   $$\frac{\csc^2 x}{\csc^2 x} - \frac{2\cot x}{\csc^2 x} = 1-2\sin x\cos x$$

5. **Simplify the first term. Rewrite the numerator and denominator in the second term by using the ratio and reciprocal identities.**

$$1 - \frac{2\dfrac{\cos x}{\sin x}}{\dfrac{1}{\sin^2 x}} = 1 - 2\sin x \cos x$$

6. **Simplify the complex fraction by flipping the denominator and multiplying it by the numerator.**

$$1 - 2\frac{\cos x}{\cancel{\sin x}} \cdot \frac{\sin^{\cancel{2}} x}{1} = 1 - 2\sin x \cos x$$

$$1 - 2\cos x \sin x = 1 - 2\sin x \cos x$$

Identifying With the Operations

Identities that have sums, differences, multiple angles, and half-angles have a suggested procedure just staring at you, because having all the functions in terms of the same angle — not twice one or the sum of the other two — is best. You just apply whatever identity the term in the equation is equal to — substitute in the equivalence of the identity — and proceed from there.

Adding it up

Sum and difference identities usually involve two different angles and then a third combined angle. To prove the identity, you need to get rid of that third angle. The first example involves a sum of two different angles.

The identity $\dfrac{\cos(x+y)}{\cos x \cos y} = 1 - \tan x \tan y$ uses the angles x and y.

1. **Replace the cosine of the sum of the two angles with its identity.**

$$\frac{\cos x \cos y - \sin x \sin y}{\cos x \cos y} = 1 - \tan x \tan y$$

2. **Break up the fraction by putting each term in the numerator over the denominator.**

$$\frac{\cos x \cos y}{\cos x \cos y} - \frac{\sin x \sin y}{\cos x \cos y} = 1 - \tan x \tan y$$

3. **Reduce the first fraction. Rewrite the second fraction as the product of two fractions. Then replace the two fractions in that product by using the ratio identity.**

$$1 - \frac{\sin x}{\cos x} \cdot \frac{\sin y}{\cos y} = 1 - \tan x \tan y$$

$$1 - \tan x \tan y = 1 - \tan x \tan y$$

The next example shows an identity for three times an angle: $\sin 3\theta = 3\sin\theta - 4\sin^3\theta$.

1. **Replace the 3θ with the sum of θ and 2θ to create the identity for the sum of two angles.**

$$\sin 3\theta = \sin(\theta + 2\theta)$$

2. **Apply the angle-sum identity for sine.**

$$\sin(\theta + 2\theta) = \sin\theta\cos 2\theta + \cos\theta\sin 2\theta$$

3. **Now replace $\cos 2\theta$ and $\sin 2\theta$ by using the double-angle identities.**

$$\sin(\theta + 2\theta) = \sin\theta(1 - 2\sin^2\theta) + \cos\theta(2\sin\theta\cos\theta)$$

You have to choose the right identity for the $\cos 2\theta$. I used the one that I did because I had an eye on what the final result is to be — all sines.

4. **Multiply through on the right side.**

$$\sin(\theta + 2\theta) = \sin\theta - 2\sin^3\theta + 2\sin\theta\cos^2\theta$$

5. **Replace $\cos^2\theta$ with its equivalent by using the Pythagorean identity. Then simplify the terms.**

$$\begin{aligned}\sin(\theta + 2\theta) &= \sin\theta - 2\sin^3\theta + 2\sin\theta(1 - \sin^2\theta)\\ &= \sin\theta - 2\sin^3\theta + 2\sin\theta - 2\sin^3\theta\\ &= 3\sin\theta - 4\sin^3\theta\end{aligned}$$

What difference does it make?

Using functions involving the difference between two angle measures has many of the same features as those with sums, so I added a couple of twists to the examples in this section.

First, solve the identity $\cot(x - 45°) = \dfrac{1 + \tan x}{\tan x - 1}$, which incorporates some function values for a 45-degree angle.

1. **Rewrite the cotangent of the difference by using the reciprocal identity, because the cotangent doesn't have a standard difference identity.**

$$\cot(x - 45°) = \frac{1}{\tan(x - 45°)}$$

2. **Now replace the tangent of the difference with its identity.**

$$\frac{1}{\tan(x - 45°)} = \frac{1}{\dfrac{\tan x - \tan 45°}{1 + \tan x\tan 45°}}$$

3. **Flip the denominator and multiply it by the numerator, which is 1.**

$$\frac{1}{\tan(x-45°)} = \frac{1+\tan x \tan 45°}{\tan x - \tan 45°}$$

4. **The value of the tangent of 45 degrees is 1, so replace all those terms with 1.**

$$\frac{1}{\tan(x-45°)} = \frac{1+\tan x(1)}{\tan x - 1} = \frac{1+\tan x}{\tan x - 1}$$

The next example proves that tan x is equal to itself. Yes, I know that seems a bit bizarre — obviously, it must be true. (As Gomer Pyle would say, "Surprise! Surprise!") Discovering the technique involved is what makes going through the steps worth the effort. The trick here is to write tan x as the tangent of the difference of the angles $2x$ and x.

1. **Write the difference identity for tangent.**

$$\tan x = \tan(2x-x) = \frac{\tan 2x - \tan x}{1+\tan 2x \tan x}$$

2. **Replace tan $2x$ with its double-angle identity.**

$$\tan x = \frac{\dfrac{2\tan x}{1-\tan^2 x} - \tan x}{1+\dfrac{2\tan x}{1-\tan^2 x} \cdot \tan x}$$

3. **Get rid of the complex fraction by multiplying every term in the numerator and denominator by $1 - \tan^2 x$.**

$$\tan x = \frac{\left(\dfrac{2\tan x}{1-\tan^2 x} - \tan x\right)(1-\tan^2 x)}{\left(1+\dfrac{2\tan x}{1-\tan^2 x} \cdot \tan x\right)(1-\tan^2 x)}$$

$$\tan x = \frac{\dfrac{2\tan x}{1-\tan^2 x}(1-\tan^2 x) - \tan x(1-\tan^2 x)}{1(1-\tan^2 x) + \dfrac{2\tan x}{1-\tan^2 x} \cdot \tan x(1-\tan^2 x)}$$

4. **Simplify the fractions.**

$$\tan x = \frac{\dfrac{2\tan x}{1-\tan^2 x}\left(1-\tan^2 x\right) - \tan x(1-\tan^2 x)}{1(1-\tan^2 x) + \dfrac{2\tan x}{1-\tan^2 x} \cdot \tan x\left(1-\tan^2 x\right)}$$

$$= \frac{2\tan x - \tan x(1-\tan^2 x)}{1(1-\tan^2 x) + 2\tan x \cdot \tan x}$$

5. **Multiply through the parentheses.**

$$\tan x = \frac{2\tan x - \tan x + \tan^3 x}{1-\tan^2 x + 2\tan^2 x}$$

6. **Combine the like terms in the numerator and denominator.**

$$\tan x = \frac{\tan x + \tan^3 x}{1 + \tan^2 x}$$

7. **Factor the numerator. Then reduce the fraction.**

$$\tan x = \frac{\tan x \left(1 + \cancel{\tan^2 x}\right)}{1 + \cancel{\tan^2 x}} = \tan x$$

Multiplying your fun

The only particular challenge involved in dealing with identities when using the multiple-angle formulas is in deciding which version of cos 2θ to use or whether to add angles together or double them. Here are some examples that illustrate these situations.

Solve the identity $\dfrac{\sin 2\theta + \cos 2\theta}{\sin \theta \cos \theta} = 2 + \cot \theta - \tan \theta$. You'll have to make a decision.

1. **Apply the double-angle identities for sin 2θ and cos 2θ.**

$$\frac{2\sin \theta \cos \theta + \cos^2 \theta - \sin^2 \theta}{\sin \theta \cos \theta} = 2 + \cot \theta - \tan \theta$$

Choosing the formula for the sine isn't a problem, because you have only one to choose from. The cosine, however, takes some observations. Because you have both sine and cosine in the denominator, you don't want to use the double-angle identities for cosine that have a 1 in them.

2. **Split up the fraction, writing each term in the numerator over the denominator.**

$$\frac{2\sin \theta \cos \theta}{\sin \theta \cos \theta} + \frac{\cos^2 \theta}{\sin \theta \cos \theta} - \frac{\sin^2 \theta}{\sin \theta \cos \theta} = 2 + \cot \theta - \tan \theta$$

3. **Reduce the fractions.**

$$\frac{2\cancel{\sin \theta}\cancel{\cos \theta}}{\cancel{\sin \theta}\cancel{\cos \theta}} + \frac{\cos^{\cancel{2}} \theta}{\sin \theta \cancel{\cos \theta}} - \frac{\sin^{\cancel{2}} \theta}{\cancel{\sin \theta} \cos \theta} = 2 + \cot \theta - \tan \theta$$

$$2 + \frac{\cos \theta}{\sin \theta} - \frac{\sin \theta}{\cos \theta} = 2 + \cot \theta - \tan \theta$$

4. **Replace the fractions with their equivalents by using the ratio identities.**

$$2 + \cot \theta - \tan \theta = 2 + \cot \theta - \tan \theta$$

Finding an identity for an angle with a multiple greater than 2 requires that you decide whether to use a sum identity or a double-angle identity. The best approach isn't always clear. Sometimes the other terms in the equation give you a hint. Often, you just flip a coin. Solve the identity $\sin 4x = 4 \sin x \cos^3 x - 4 \sin^3 x \cos x$.

1. **Rewrite 4x as 2(2x).**

   $$\sin 4x = \sin 2\,(2x)$$

2. **Insert this new angle into the double-angle identity.**

   $$\sin 2\,(2x) = 2 \sin 2x \cos 2x$$

3. **Replace sin 2x and cos 2x with the double-angle identities, choosing the best cosine identity for the situation.**

   In this case, because you have a difference of terms involving third-degree powers of sine and cosine, the cosine identity involving second-degree powers of sine and cosine seems to be a good choice.

   $$\sin 2\,(2x) = 2\,(2 \sin x \cos x)(\cos^2 x - \sin^2 x)$$

4. **Multiply through.**

   $$\sin 2\,(2x) = 4 \sin x \cos^3 x - 4 \sin^3 x \cos x = \sin 4x$$

Halving fun, wish you were here

The last type of identity that you can incorporate into solving an identity or doing identity problems is the half-angle identity. This identity actually comes in very handy in calculus — not by changing from a half-angle identity to angles that are larger, but by changing from larger angles to half-angle identities. The examples in this section show some of the possibilities.

Solve the identity $\tan \dfrac{\theta}{2} = \csc \theta - \cot \theta$ by using the half-angle identity for tangent. Neither side looks very promising for solving the identity until you notice that you have two different angles — one half the size of the other. You need to apply the half-angle identity to get everything in terms of the same angle.

1. **Substitute in the identity for the half-angle of tangent.**

   $$\tan \frac{\theta}{2} = \frac{\sin \theta}{1 + \cos \theta} = \frac{1 - \cos \theta}{\sin \theta}$$

   You have two different versions to choose between. The easier one to work with is the one with two terms in the numerator: $\dfrac{1 - \cos \theta}{\sin \theta} = \csc \theta - \cot \theta$.

2. **Split up the fraction on the left by putting each term in the numerator over the denominator.**

$$\frac{1}{\sin\theta} - \frac{\cos\theta}{\sin\theta} = \csc\theta - \cot\theta$$

3. **Replace the two terms on the left with their reciprocal and ratio identities.**

$$\csc\theta - \cot\theta = \csc\theta - \cot\theta$$

This last example incorporates the half-angle of the tangent as well as the half-angle of a reciprocal function. Solve the identity $\sin\theta\sec^2\frac{\theta}{2} = 2\tan\frac{\theta}{2}$.

1. **Use the reciprocal of the half-angle identity for cosine to replace the half-angle of secant.**

   Hold off on deciding which version of the tangent's half-angle formula to use until you see what you need.

   $$\sin\theta\left(\frac{1}{\cos\frac{\theta}{2}}\right)^2 = 2\tan\frac{\theta}{2}$$

   $$\sin\theta\left(\frac{1}{\sqrt{\frac{1+\cos\theta}{2}}}\right)^2 = 2\tan\frac{\theta}{2}$$

2. **Flip the fraction in the parentheses and square the radical, leaving the terms with no radical.**

   $$\sin\theta\left(\frac{2}{1+\cos\theta}\right) = 2\tan\frac{\theta}{2}$$

3. **Choose the half-angle tangent identity that matches what you have on the left and simplify.**

   $$\sin\theta\left(\frac{2}{1+\cos\theta}\right) = 2\left(\frac{\sin\theta}{1+\cos\theta}\right)$$

   $$\left(\frac{2\sin\theta}{1+\cos\theta}\right) = \left(\frac{2\sin\theta}{1+\cos\theta}\right)$$

Part IV
Equations and Applications

The 5th Wave   By Rich Tennant

"WHAT EXACTLY ARE WE SAYING HERE?"

In this part . . .

Trig *identities* work for all values — you can put in any number in place of the variable and they're true. Not so with trig equations, though, which usually have a bunch of very selective solutions. These equations are the topic of this important part. Here, you discover inverse trig functions, which are very useful when solving trig equations. And then you run smack into the face of the law: the law of sines and law of cosines, that is.

Chapter 13

Investigating Inverse Trig Functions

As thrilling and fulfilling as the original six trig functions are, they just aren't complete without their inverses. An *inverse trig function* behaves like the inverse of any other type of function — it undoes what the original function did. Functions can have inverses if they're one-to-one, meaning each output value occurs only once. This whole inverse idea is going to take some fast talking when it comes to trig functions, because they keep repeating values over and over as angles are formed with every full rotation of the circle — so you're going to wonder how these functions and inverses can be one-to-one. If you need a refresher on basic inverse functions, flip on back to Chapter 3 for the lowdown on them and how you determine one. This chapter, on the other hand, takes you deeper into the world of inverses and shows how the inverses of *trig* functions work. You'll also see why you'd even want to bother.

Writing It Right

You use inverse trig functions to solve equations such as $\sin x = \frac{1}{2}$, $\sec x = -2$, or $\tan 2x = 1$. In typical algebra equations, you can solve for the value of x by dividing each side of the equation by the coefficient or by adding the same thing to each side, and so on. But you can't do that with the function $\sin x = \frac{1}{2}$. Would it make sense to divide each side by *sine?* "Out, out thou sine!" Here's what you'd get: $\frac{\sin x}{\sin} = \frac{\frac{1}{2}}{\sin}$, $x = \frac{\frac{1}{2}}{\sin}$. Goodness, no! That's silly.

Using the notation

To find the inverse of an equation such as $\sin x = \frac{1}{2}$, solve for the following statement: "x is equal to the angle whose sine is $\frac{1}{2}$." In trig speak, you write this statement as $x = \sin^{-1}\left(\frac{1}{2}\right)$. The notation involves putting a –1 in the superscript position. Here are some more examples of trig equations with their corresponding inverses and the translation.

Function	Inverse	What It Means
$\sec x = -2$	$x = \sec^{-1}(-2)$	x is the angle whose secant is –2
$\tan 2x = 1$	$2x = \tan^{-1}(1)$	$2x$ is the angle whose tangent is 1
$\cos\theta = 0$	$\theta = \cos^{-1}(0)$	θ is the angle whose cosine is 0
$\csc\alpha = -1$	$\alpha = \csc^{-1}(-1)$	α is the angle whose cosecant is –1

Interpreting the exponent

You've undoubtedly seen and used the exponent –1 in math expressions before now. But that exponent does a different kind of job for inverse trig functions and relations. The notation for an inverse trig relation such as $\sin^{-1}x$ means that you want an *inverse* for the expression, not the *reciprocal*. (If you want $\frac{1}{\tan x}$, you have to write $(\tan x)^{-1}$ with parentheses.) The –1 exponent is where the exponential notation for trig functions makes a big exception.

When raising trig functions to a power, $\sin^2 x = (\sin x)^2$ and $\cos^4 x = (\cos x)^4$, but $\tan^{-1}x$ means the inverse function, not raising $\tan x$ to the –1 power.

Alternating the notation

Inverses of trig functions have an alternate notation that avoids the confusion over what the –1 superscript means: the *arc* name. Another way of saying $\sin^{-1}x$ is *arcsin x*. The inverse cosine is $\cos^{-1}x$, or *arccos x*. The other inverse functions are *arctan x, arccsc x, arcsec x,* and *arccot x*. This notation is longer to write out and is sometimes awkward to write, so the original superscript notation is often preferable. You'll see inverse functions written both ways, though.

Triskaidekaphobia

Are you a triskaidekaphobe? If so, you're in good company. Some famous triskaidekaphobes include Napoleon, Mark Twain, and presidents Herbert Hoover and Franklin Roosevelt. *Triskaidekaphobia* is the fear or avoidance of the number 13. To many, the number 13 means bad luck, which is why some hotels skip the number 13 when they're numbering the floors — too many people don't want to stay on that floor.

Where did this fear of the number 13 start? Many theories exist. Some say that it began in medieval Christian countries, because 13 people were at the Last Supper of Christ. The number 13 was also unlucky in Norse mythology. According to legend, an evil Norse god went uninvited to a party for 12 at Valhalla and caused the death of the god of light, joy, and reconciliation.

A relatively recent disaster is linked to the number 13: the famous Apollo 13 space mission. This flight was launched on April 11, 1970. Now, are you ready for this? If you add up 4 (for the month), 11, and 70 (for the year), you get 85. And the sum of the digits in 85 equals 13. Also, the flight was launched from pad 39 (which is 13 times 3) at 13:13 local time and was struck by an explosion on April 13th.

To top it off, if you add a Friday to the mix and worry about Friday the 13th, you have several days a year that you need to stay in the house — or not. Have you noticed which chapter this is?

Distinguishing between the few and many

Technically, an inverse function is supposed to have only one answer. (Part of the definition of an inverse is that the function and inverse are one-to-one.) Each input has one output, and each output has one input. For all the practical uses of trig inverses, you have a way around this rule. You can designate whether you want one answer or many answers by using either the inverse *function* or the inverse *relation*. A relation is a bit looser than a function; it allows more than one output to have the same input. To differentiate between these two entities, the common practice is to use capital letters for the function and lowercase letters for the relation.

Trig Functions	*Trig Relations*
$\text{Sin}^{-1} x$ or $\text{Arcsin} \, x$	$\sin^{-1} x$ or $\arcsin x$
$\text{Cos}^{-1} x$ or $\text{Arccos} \, x$	$\cos^{-1} x$ or $\arccos x$
$\text{Tan}^{-1} x$ or $\text{Arctan} \, x$	$\tan^{-1} x$ or $\arctan x$
$\text{Cot}^{-1} x$ or $\text{Arccot} \, x$	$\cot^{-1} x$ or $\text{arccot} \, x$
$\text{Sec}^{-1} x$ or $\text{Arcsec} \, x$	$\sec^{-1} x$ or $\text{arcsec} \, x$
$\text{Csc}^{-1} x$ or $\text{Arccsc} \, x$	$\csc^{-1} x$ or $\text{arcscs} \, x$

If you write the function $\text{Sin}^{-1}\left(\frac{1}{2}\right)$, the result is 30 degrees (or $\frac{\pi}{6}$ radians). Just one answer exists, which is called the *principal value* of the inverse. But if you write $\sin^{-1}\left(\frac{1}{2}\right)$, then the result can be 30 degrees, 150 degrees, 390 degrees, 510 degrees, and so on (or $\frac{\pi}{6}, \frac{5\pi}{6}, \frac{13\pi}{6}, \frac{17\pi}{6}, \frac{25\pi}{6}, \frac{29\pi}{6} \ldots$). It all depends on the situation — what you want at the time. Do you want just the principal value, or do you want multiple values? Or you may want a bunch of values within one full rotation — from 0 to 360 degrees.

When you want lots and lots of angles or answers, listing them all can be tedious. In fact, listing every possible solution may not even be doable. Rather than making a list, you can give a *rule,* which allows you to define an angle with all its full-rotation multiples — the angles with the same terminal side.

Let n represent any integer ($\ldots, -3, -2, -1, 0, 1, 2, 3, \ldots$). Using the n as a multiplier, you can write a long list of angles more efficiently. Instead of saying $x = $ 30, 150, 390, 510, 750, 870, $\ldots$, divide the list into two groups: $x = $ 30; 390; 750; 1110; $\ldots$; and $x = $ 150; 510; 870; 1230; $\ldots$; and then use the two rules that follow:

$$x = 30 + 360n \text{ or } x = 150 + 360n$$

And then, in radians, instead of saying $x = \frac{\pi}{6}, \frac{13\pi}{6}, \frac{25\pi}{6} \ldots$ or $x = \frac{5\pi}{6}, \frac{17\pi}{6}, \frac{29\pi}{6} \ldots$, use these two rules:

$$\frac{\pi}{6} + 2\pi n \text{ or } \frac{5\pi}{6} + 2\pi n$$

Here's an example showing how to write all the angles that have a cosine equal to $\frac{1}{2}$. The steps involve solving the inverse relation, not just finding the principal value for the function. Solve for the values that satisfy $\cos^{-1}\left(\frac{1}{2}\right)$.

1. **List several solutions in both degrees and radians.**

   $$\cos^{-1}\left(\frac{1}{2}\right) = 60°, 300°, 420°, 660°, 780°, 1020° \ldots$$

   $$\cos^{-1}\left(\frac{1}{2}\right) = \frac{\pi}{3}, \frac{5\pi}{3}, \frac{7\pi}{3}, \frac{11\pi}{3}, \frac{13\pi}{3}, \frac{17\pi}{3} \ldots$$

2. **Write the answers in degrees by using the first two angles plus multiples of 360.**

   $$\cos^{-1}\left(\frac{1}{2}\right) = 60° + 360°n \text{ or } 300° + 360°n$$

3. **Write the answers in radians by using the first two angles plus multiples of 2π.**

   $$\cos^{-1}\left(\frac{1}{2}\right) = \frac{\pi}{3} + 2\pi n \text{ or } \frac{5\pi}{3} + 2\pi n$$

Writing all the possible angles for inverse tangent is a bit easier than for sine or cosine. The tangent is positive in the first and third quadrants, which are cattycorner from one another (half a full rotation). Because of this fact, the angles that have the same function values are 180 degrees apart, and you can use nice multiples of 180 degrees or π to name all the answers. This isn't the case with sine and cosine, though. The angles with the same function values are in quadrants that are adjacent to one another, so you have to use two separate rules — both with multiples of 360 degrees — to name all the answers.

Here's how to write all the angles that have a tangent equal to $-\dfrac{\sqrt{3}}{3}$. Solve for values that satisfy $\tan^{-1}\left(-\dfrac{\sqrt{3}}{3}\right)$:

1. **List several answers in both degrees and radians.**

$$\tan^{-1}\left(-\frac{\sqrt{3}}{3}\right) = 150°, 330°, 510°, 690° \ldots$$

$$\tan^{-1}\left(-\frac{\sqrt{3}}{3}\right) = \frac{5\pi}{6}, \frac{11\pi}{6}, \frac{17\pi}{6}, \frac{23\pi}{6} \ldots$$

2. **Write the answers in degrees by using multiples of 180.**

$$\tan^{-1}\left(-\frac{\sqrt{3}}{3}\right) = 150° + 180°n$$

3. **Write the answers in radians by using multiples of π.**

$$\tan^{-1}\left(-\frac{\sqrt{3}}{3}\right) = \frac{5\pi}{6} + \pi n$$

Determining Domain and Range of Inverse Trig Functions

A function that has an inverse has exactly one output (belonging to the *range*) for every input (belonging to the *domain*), and vice versa. To keep inverse trig functions consistent with this definition, you have to designate ranges for them that will take care of all the possible input values and not have any duplication. The output values of the inverse trig functions are all angles — in either degrees or radians — and they're the answer to the question, "Which angle gives me this number?" In general, the output angles for the individual inverse functions are paired up as angles in Quadrants I and II or angles in Quadrants I and IV. The quadrants are selected this way for the inverse trig functions because the pairs are adjacent quadrants, allowing for both positive and negative entries. The notation for these inverse functions uses capital letters (see the preceding section).

Inverse sine function

The domain for $\text{Sin}^{-1} x$, or Arcsin x, is from –1 to 1. In mathematical notation, the domain or input values, the x's, fit into the expression $-1 \leq x \leq 1$, because no matter what angle measure you put into the sine function, the output is restricted to these values. The range, or output, for $\text{Sin}^{-1} x$ is all angles from –90 to 90 degrees or, in radians, $-\frac{\pi}{2}$ to $\frac{\pi}{2}$. If the output is the angle θ, then you write these expressions as $-90° \leq \theta \leq 90°$ or $-\frac{\pi}{2} \leq \theta \leq \frac{\pi}{2}$. The inputs are angles in the adjacent Quadrants I and IV, because the sine is positive in the first quadrant and negative in the second quadrant. Those angles cover all the possible input values.

Inverse cosine function

The domain for $\text{Cos}^{-1} x$, or Arccos x, is from –1 to 1, just like the inverse sine function. So the x (or input) values are $-1 \leq x \leq 1$. The range for $\text{Cos}^{-1} x$ consists of all angles from 0 to 180 degrees or, in radians, 0 to π. If the output is the angle θ, then you write these expressions as $0° \leq \theta \leq 180°$ or $0 \leq \theta \leq \pi$. The inputs are angles in the adjacent Quadrants I and II, because the cosine is positive in the first quadrant and negative in the second quadrant. Those angles cover all the possible input values for the function.

Inverse tangent function

The domain for $\text{Tan}^{-1} x$, or Arctan x, is all real numbers — numbers from $-\infty$ to ∞. This is because the output of the tangent function, this function's inverse, includes all numbers. The range, or output, of $\text{Tan}^{-1} x$ is angles *between* –90 and 90 degrees or, in radians, *between* $-\frac{\pi}{2}$ and $\frac{\pi}{2}$. One important note is that the range doesn't include those beginning and ending angles; the tangent function isn't defined for –90 or 90 degrees. The range of $\text{Tan}^{-1} x$ includes all the angles in the adjacent Quadrants I and IV, except for the two angles with terminal sides on the y-axis.

Inverse cotangent function

The domain of $\text{Cot}^{-1} x$, or Arccot x, is the same as that of the inverse tangent function. The domain includes all real numbers. The range, though, is different — it includes all angles *between* 0 and 180 degrees (*between* 0 and π). So any angle in Quadrants I and II is included in the range, except for those with terminal sides on the x-axis. Those two angles aren't in the domain of the cotangent function (see Chapter 8), so they aren't in the range of the inverse.

Inverse secant function

The domain of $\text{Sec}^{-1} x$, or Arcsec x, consists of all the numbers from 1 on up plus all the numbers from −1 on down. Letting x be the input, you write this expression as $x \geq 1$ or $x \leq -1$. In other words, the domain includes all the numbers from $-\infty$ to ∞, except for the numbers between −1 and 1. The range of $\text{Sec}^{-1} x$ is all the angles *between* 0 and 180 degrees (*between* 0 and π) — meaning all angles in Quadrants I and II, with the exception of 90 degrees, or $\frac{\pi}{2}$.

Inverse cosecant function

The domain of $\text{Csc}^{-1} x$, or Arccsc x, is the same as that for the inverse secant function, all the numbers from 1 on up plus all the numbers from −1 on down. The range is different, though — it includes all angles *between* −90 and 90 degrees or, in radians, *between* $-\frac{\pi}{2}$ and $\frac{\pi}{2}$. In short, the range is all the angles in the Quadrants I and IV, with the exception of 0 degrees, or 0 radians.

Summarizing domain and range

Sometimes, looking at a chart or summary of the domains and ranges of the inverse trig functions is more informative than reading about them. Take a look at Table 13-1. You should notice some patterns — some similarities and differences. The ranges of three of the functions are in Quadrants I and II, and the other three are in Quadrants I and IV. The reciprocals sine and cosecant use the same quadrants. So do the reciprocals cosine and secant. The tangent and cotangent don't use the same quadrants, though.

Table 13-1	Domains and Ranges of the Inverse Trig Functions		
Inverse Trig Function	**Domain**	**Range**	**Quadrants in Range**
$\text{Sin}^{-1} x$	$-1 \leq x \leq 1$	$-90° \leq \theta \leq 90°$	I and IV
$\text{Cos}^{-1} x$	$-1 \leq x \leq 1$	$0° \leq \theta \leq 180°$	I and II
$\text{Tan}^{-1} x$	$-\infty < x < \infty$	$-90° < \theta < 90°$	I and IV
$\text{Cot}^{-1} x$	$-\infty < x < \infty$	$0° < \theta < 180°$	I and II
$\text{Sec}^{-1} x$	$x \geq 1$ or $x \leq -1$	$0° < \theta < 180°$, $\theta \neq 90°$	I and II
$\text{Csc}^{-1} x$	$x \geq 1$ or $x \leq -1$	$-90° < \theta < 90°$, $\theta \neq 0°$	I and IV

Triangle dissection paradox

Look at the two triangles in the following figure, made up of the same four pieces. They're sitting on the same grid, appearing to take up the same space, but one has a hole or gap. How can that be?

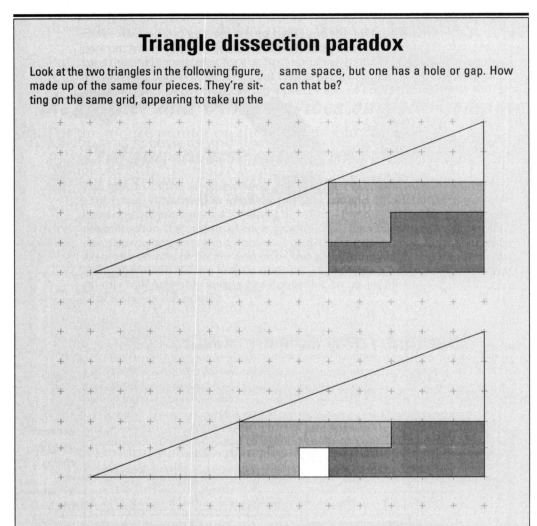

This situation happens because the hypotenuse of the main outer triangle isn't really a straight line. The slopes of the sides of the two triangles (the hypotenuses) making up that larger hypotenuse aren't the same; the hypotenuse of the top triangle is slightly bent in, and the hypotenuse of the bottom triangle is slightly bent out. Of course, you're not supposed to be able to notice this bend, but you can check it by laying a ruler or straightedge along the hypotenuse. You might call this figure an optical illusion.

Working with Inverses

The easiest way to work with inverse trig functions is to have a chart handy with the exact values of the functions, which you can find in the Appendix. When angles other than the most common or popular are involved, you can either use a table such as the one in the Appendix or get out your handy-dandy scientific calculator.

This first example uses the exact value from a chart. Find $\text{Cos}^{-1}\left(-\dfrac{\sqrt{2}}{2}\right)$.

1. **Determine the reference angle that you need by using the absolute value of the input.**

   The value $\dfrac{\sqrt{2}}{2}$ is the cosine of 45 degrees or $\dfrac{\pi}{4}$ radians.

2. **Use the sign of the input to determine the correct quadrant.**

   Because $-\dfrac{\sqrt{2}}{2}$ is negative, and of the two quadrants for the range (refer to Table 13-1), the cosine is negative in QII, then the answer is an angle in QII whose reference angle is 45 degrees.

3. **Determine the correct angle measure.**

   The angle in standard position in QII whose reference angle is 45 degrees is $180 - 45 = 135$ or $\pi - \dfrac{\pi}{4} = \dfrac{3\pi}{4}$. So $\text{Cos}^{-1}\left(-\dfrac{\sqrt{2}}{2}\right) = 135°$ or $\dfrac{3\pi}{4}$. Refer to Chapter 8 for more on reference angles.

The next example involves inverse cotangent. Find $\text{Cot}^{-1}\left(\sqrt{3}\right)$.

1. **Determine the reference angle that you need.**

   The value $\sqrt{3}$ is the cotangent of 30 degrees or $\dfrac{\pi}{6}$.

2. **Use the sign of the input to determine the correct quadrant.**

   Because $\sqrt{3}$ is positive in QI, then the answer is the angle in QI whose reference angle is $30°$ or $\dfrac{\pi}{6}$.

3. **Determine the correct angle measure.**

   All angles in QI are the same as their reference angles, so $\text{Cot}^{-1}\left(\sqrt{3}\right) = 30°$ or $\dfrac{\pi}{6}$.

The problems that you encounter won't always involve nice numbers from the most common angles. When you're faced with a nasty little decimal value, you may have to use a table. In this next example, you start off with a decimal value, and an answer to the nearer *degree* is the proper response. The decimal in the following example is rounded off to three decimal places. To do these problems, you find the closest answer.

Find Arctan(−3.732).

1. **Determine the reference angle that you need.**

   Use the Appendix, and you can see that the value 3.732 corresponds to the tangent of a 75-degree angle. This angle is the closest in whole degrees to having a tangent of 3.732.

2. **Use the sign of the input to determine the correct quadrant.**

   Because −3.732 is negative, the answer is an angle in QIV whose reference angle is 75 degrees.

3. **Determine the correct angle measure.**

   In QIV, a reference angle of 75 degrees has a measure of either −75 degrees or its positive equivalent (same terminal side), 285 degrees. So Arctan(−3.732) = Tan^{-1}(−3.732) = −75° or 285°.

Getting Friendly with Your Calculator

Scientific calculators are wonderful tools — they make life easier and improve the quality (correctness) of the results. In most instances, computing inverse trig functions with a calculator is quick and easy. You need to be aware of a couple of pitfalls, though.

If you've already trekked to the store and left empty-handed and bewildered at the array of graphing calculators, your best bet is to hold your head high; confidently head back there; and pick up a model by Texas Instruments, Sharp, or Hewlett-Packard — those models seem to be the most popular. (For what it's worth, I'm a Texas Instruments gal — especially when it comes to their graphing calculators.)

Changing the mode

Scientific calculators are very accommodating — they give you results in either degrees or radians, depending on which *mode* you set them in. This feature is great, but it trips up even the best mathematicians from time to time. Each calculator is different, but they always have either a single button or a multiple-button sequence that switches from radians to degrees and back again — sort of like a toggle switch. Some calculators even have a legend at the top or bottom of the screen that tells you whether you're in degree or radian mode, perhaps as obvious as an *R* or *D*. Problems tend to arise when you use your calculator for more than one task. Perhaps you're studying both trigonometry and physics, and one calls for degree mode while the other calls for radian mode. Just be aware, and you won't get caught.

Interpreting notation on the calculator

Calculator notation, or the mumbo jumbo on the buttons, is somewhat tricky. Even though the –1 superscript indicates an inverse trig function when written in a book or on your paper, you can't use the –1 button to find the value of an inverse trig function on a calculator. On scientific calculators, the –1 or x^{-1} button means to find the reciprocal of a number. Look under the *2nd functions,* which are different functions or operations written above the buttons, for the inverse trig functions. They're usually above the original sine, cosine, and tangent buttons. Some calculators have a button labeled "2nd." Others use alternate colors — usually yellow or green — to denote the second use of the button. And even when you find the inverse functions, you'll notice that they're only for the three primary trig functions. The calculator doesn't show any for cosecant, secant, or cotangent. So where are they? First, I discuss how to use the three buttons available and then I tell you how to calculate the other inverses.

Using the inverse function button

To explain this button, I use an example. Here's how you find $\mathrm{Sin}^{-1}\left(\frac{1}{2}\right)$ in degrees by using a scientific calculator.

1. **Decide whether you want your answer in radians or degrees.**

   For this example, use the *mode* menu or whatever method your calculator uses to change the mode to degrees.

2. **Enter the problem as given.**

   The following are the typical keystrokes: 2nd Sin .5 Enter . The result is 30, meaning 30 degrees.

Calculating the inverse of a reciprocal function

To determine the inverse of a reciprocal function, such as $\mathrm{Cot}^{-1}(2)$ or $\mathrm{Sec}^{-1}(-1)$, you have to change the problem back to the function's reciprocal — one of the three basic functions — and then use the appropriate inverse button.

What would Babe and Hank think?

Mathematicians are always looking for patterns and relationships among numbers. When someone makes a discovery, they get to give the new pattern or relationship a name, which is why some pairs of numbers are called *Ruth-Aaron Pairs.*

When two consecutive numbers have the sums of their *prime factors* (all the prime numbers whose product is that number) equal in value, they're called a *Ruth-Aaron Pair.* The pair of numbers that inspired this name is 714 and 715. The prime factors of 714 are 2, 3, 7, and 17. The

prime factors of 715 are 5, 11, and 13. The sums are 2 + 3 + 7 + 17 = 29 and 5 + 11 + 13 = 29. These two numbers correspond to Hank Aaron's record-breaking 715th home run, which he hit in 1974 to break Babe Ruth's record of 714 home runs.

Some other Ruth-Aaron Pairs are 5 and 6, 24 and 25, 49 and 50, and 77 and 78. These particular pairs work if you only count multiple factors once. The number 24 has a factor of 2 three times, but you only count it once — and so it goes with 25, which has two factors of 5.

When changing to the function's reciprocal, you flip the number with that function, too. For example, $\mathrm{Cot}^{-1}(2)$ becomes $\mathrm{Tan}^{-1}\left(\frac{1}{2}\right)$. You change $\mathrm{Arcsec}\left(-\frac{2}{\sqrt{3}}\right)$ to $\mathrm{Arccos}\left(-\frac{\sqrt{3}}{2}\right)$, $\mathrm{Csc}^{-1}(1)$ to $\mathrm{Sin}^{-1}(1)$, and so on.

Working around the inverse cotangent

The other big pitfall you encounter when using a scientific calculator involves the inverse cotangent. The inverse tangent, $\mathrm{Tan}^{-1}x$, has its range in QI and QIV, but $\mathrm{Cot}^{-1}x$ has its range in QI and QII. If you want $\mathrm{Cot}^{-1}\left(-\sqrt{3}\right)$, for example, and you use $\mathrm{Tan}^{-1}\left(-\frac{1}{\sqrt{3}}\right)$ and your calculator, you get an answer in the fourth quadrant. You have to be aware that this quadrant isn't correct; you got it because you changed functions so you could use the calculator. Just use the answer from the calculator and determine the corresponding angle in QII. Here's an example.

Find $\mathrm{Cot}^{-1}\left(-\sqrt{3}\right)$ in degrees.

1. **Set the mode to degrees.**

2. **Change the function and value to their reciprocals.**

   $\mathrm{Cot}^{-1}\left(-\sqrt{3}\right)$ becomes $\mathrm{Tan}^{-1}\left(-\frac{1}{\sqrt{3}}\right)$

3. **Find the value of the inverse function by using a calculator.**

   Enter 2nd Tan (−) 1/ √3 Enter. On some calculators, parentheses will automatically pop up for you to enter the tangent value inside them. If they don't, then you should insert parentheses around the fraction your-self. The result is –30 degrees.

4. **Find the angle in QII that has the same reference angle.**

   The angle in QII with a 30-degree reference angle is an angle of 150 degrees.

Multiplying the Input

Multiple-angle functions are those such as sin 2θ, cos $3x$, tan 6β, and so on. When considering inverse relations (which give multiple answers) for these angles, the multiplier usually indicates how many more times as many answers that problem has as compared to a similar problem without a multiplier. For example, the equation $\sin\theta = \dfrac{\sqrt{3}}{2}$ has two different answers if you consider all the angles between 0 and 360 degrees: θ equals 60 and 120 degrees. But if you change the equation to $\sin 2\theta = \dfrac{\sqrt{3}}{2}$, you get twice as many, or four, answers between 0 and 360 degrees: θ equals 30, 60, 210, and 240 degrees. These angles are all within one rotation, but putting them into the original equation and multiplying gives angles with the same terminal side as the angles within one rotation.

Here are some examples to show you how this multiplication works and how to find the answers. First, I show how I got the answers to $\sin 2\theta = \dfrac{\sqrt{3}}{2}$ for $0° \leq \theta \leq 360°$.

1. **Write the inverse equation.**

   $$2\theta = \sin^{-1}\left(\frac{\sqrt{3}}{2}\right)$$

2. **List all the angles in *two* rotations, $0° \leq \theta \leq 720°$, that have that sine, and set them equal to 2θ.**

   $$2\theta = 60°, 120°, 420°, 480°$$

   The second two angles are just 360 more than the first two.

3. **Divide the terms on both sides of the equation by 2 to solve for θ.**

   $$\frac{2\theta}{2} = \frac{60°}{2}, \frac{120°}{2}, \frac{420°}{2}, \frac{480°}{2}$$
   $$\theta = 30°, 60°, 210°, 240°$$

   Notice how all the solutions for θ are between 0 and 360 degrees — just as asked.

Now solve $\cos 3x = -\dfrac{\sqrt{2}}{2}$ for any x such that $0 \le x \le 2\pi$.

1. **Write the inverse equation.**

   $$3x = \cos^{-1}\left(-\frac{\sqrt{2}}{2}\right)$$

2. **List all the angles in *three* rotations, $0 \le x \le 6\pi$, that have that cosine, and set them equal to $3x$.**

   $$3x = \frac{3\pi}{4}, \frac{5\pi}{4}, \frac{11\pi}{4}, \frac{13\pi}{4}, \frac{19\pi}{4}, \frac{21\pi}{4}$$

   The second two angles are just 2π greater than the first two, and the last two are 2π greater than the second two.

3. **Multiply both sides by $\dfrac{1}{3}$ to solve for x.**

   $$\frac{1}{3} \cdot 3x = \frac{1}{3} \cdot \frac{\cancel{3}\pi}{4}, \frac{1}{3} \cdot \frac{5\pi}{4}, \frac{1}{3} \cdot \frac{11\pi}{4}, \frac{1}{3} \cdot \frac{13\pi}{4}, \frac{1}{3} \cdot \frac{19\pi}{4}, \frac{1}{\cancel{3}} \cdot \frac{\cancel{21}^{\,7}\pi}{4}$$

   $$x = \frac{\pi}{4}, \frac{5\pi}{12}, \frac{11\pi}{12}, \frac{13\pi}{12}, \frac{19\pi}{12}, \frac{7\pi}{4}$$

   This result shows the big advantage of radians — the numbers don't get as big as they do with degrees. The disadvantage may be having so many fractions.

Solving Some Mixed Problems

When working with inverse trig functions, it's always more convenient when the numbers you're working with are the results of applying one of the trig functions to a common angle measure. When the angle isn't a common one, though, you need a calculator or table.

By using inverse trig functions, you can solve some interesting problems, where you never even need to know the angle measure. You just need to know a function value, a quadrant, and a few trig identities. For example, you can find $\cos\left(\sin^{-1}\left(\dfrac{12}{13}\right)\right)$, which says to "find the cosine of an angle whose sine is equal to $\dfrac{12}{13}$." You don't need to know the angle measure to solve this problem, but you do need to know the quadrant that the terminal side lies in, because otherwise, two different angles can be correct answers. The sine is positive in Quadrants I and II, so this problem could involve an angle in either of those quadrants, but cosine isn't positive in both of those quadrants. Consider the following example.

Find $\cos\left(\sin^{-1}\left(\dfrac{12}{13}\right)\right)$ if the terminal side of the angle is in QII.

1. **Use the Pythagorean identity to find the cosine of the angle.**

    Put the value in for sin θ, get the cosine term alone, and then take the square root of both sides.

    $$\sin^2\theta + \cos^2\theta = 1$$

    $$\left(\frac{12}{13}\right)^2 + \cos^2\theta = 1$$

    $$\cos^2\theta = 1 - \frac{144}{169} = \frac{25}{169}$$

    $$\cos\theta = \pm\sqrt{\frac{25}{169}} = \pm\frac{5}{13}$$

2. **Choose the sign of the answer.**

    Because the angle's terminal side is in QII, and the cosine is negative there, the answer is $\cos\left(\sin^{-1}\left(\frac{12}{13}\right)\right) = -\frac{5}{13}$.

The quadrant isn't a mystery in a problem that uses the inverse trig *function*. To find $\tan\left(\mathrm{Cos}^{-1}\left(-\frac{11}{61}\right)\right)$, you can assume that the angle has its terminal side in QII, because the inverse cosine function is negative in that quadrant.

1. **Use the reciprocal identity and reciprocal of the number to find the secant.**

    The problem involves the angle whose cosine is $-\frac{11}{61}$. I call that unknown angle θ and rewrite the expression in terms of the cosine of θ with that measure. I write the expression this way in order to change from an inverse trig function to a trig function so I can use the identity.

    If $\mathrm{Cos}\,\theta = -\frac{11}{61}$, then $\mathrm{Sec}\,\theta = -\frac{61}{11}$.

2. **Use the Pythagorean identity to solve for the tangent.**

    $$\tan^2\theta + 1 = \sec^2\theta$$

    $$\tan^2\theta + 1 = \left(-\frac{61}{11}\right)^2$$

    $$\tan^2\theta + 1 = \frac{3721}{121}$$

    $$\tan^2\theta = \frac{3721}{121} - 1 = \frac{3600}{121}$$

    $$\tan\theta = \pm\sqrt{\frac{3600}{121}} = \pm\frac{60}{11}$$

3. **Choose the sign of the answer.**

    Because the terminal side is in QII and the tangent is negative in that quadrant, $\tan\left(\mathrm{Cos}^{-1}\left(-\frac{11}{61}\right)\right) = -\frac{60}{11}$.

Chapter 14

Solving Trig Equations

Solving equations involving trigonometric expressions takes a pinch of this, a dab of that, a gentle stirring, and just the right temperature. No, this book isn't for a cooking class, but solving these equations requires the proper preparation and some skill — just like a successful dish.

Trig equations aren't identities. An identity is true for any angle in the domain of the function involved. A trig *equation* is true for some specific angles or input — if the equation has a solution at all.

Some trig equations require factoring skills from algebra or even the quadratic formula. Successfully solving most of them involves incorporating trig identities at the proper time. All the equations require knowledge of the function values and how inverse trig functions work (so head on back to Chapters 8 and 13 if you need a refresher). For the equation-solving enthusiasts, this chapter is where all the concepts come together for maximum fun and challenge.

The methods and techniques that you see in this chapter are those that people most frequently use to solve trig equations. A few more ways exist, but they don't come up as often. Also, you usually have more than one way of solving a particular trig equation. Your goal should always be to do it as quickly and efficiently as possible, but don't be alarmed if you seem to take the long way around. Sometimes that way just seems to make more sense. If a particular method works for you, go for it.

Generating Simple Solutions

The simplest type of trig equation is the one that you can immediately rewrite as an inverse in order to determine the solutions. Some examples of these types of equations include $\cos x = 1$, $2 \sin x + 1 = 0$, and $\cot x - \sqrt{3} = 0$. Here's how to solve them.

To solve $\cos x = 1$:

1. **Rewrite the equation as an inverse function equation.**

   $x = \cos^{-1}(1)$

2. **List the solutions for values of x when $0° \leq x < 360°$.**

   $x = 0°$

   The only time that the cosine is equal to 1 is when the angle, or input, is 0 degrees. The terminal sides of angles of 0 and 360 degrees are the same, so you don't have to list the angle measure twice.

3. **List all the solutions in general.**

   $x = 0° + 360° n$

Steps 2 and 3 illustrate the different ways that you can write the answers: either as a few within a certain interval, or as all that are possible, with a rule to describe them.

Now solve $2 \sin x + 1 = 0$ only for values of x such that $0 \leq x < 2\pi$:

1. **Rewrite the equation as an inverse function equation.**

   First, subtract 1 from each side; then divide each side by 2.

   $2 \sin x = -1$

   $\sin x = -\dfrac{1}{2}$

   $x = \sin^{-1}\left(-\dfrac{1}{2}\right)$

2. **List the solutions. Use the chart in the Appendix to find the angles that work.**

   $x = \dfrac{7\pi}{6}, \dfrac{11\pi}{6}$

This last example involves a reciprocal function. Your best bet is to begin by using a reciprocal identity and changing the problem.

Solve the equation $\cot x - \sqrt{3} = 0$ for all the values of x, in radians, that satisfy it:

1. **Solve for the trig function by adding the radical value to each side.**

   $\cot x = \sqrt{3}$

2. **Use the reciprocal identity and the reciprocal of the number to change to the tangent function and then multiply both parts of the fraction by the denominator to get rid of the radical.**

   $\tan x = \dfrac{1}{\sqrt{3}} = \dfrac{\sqrt{3}}{3}$

3. **Rewrite the equation as an inverse function equation.**

   $x = \tan^{-1}\left(\dfrac{\sqrt{3}}{3}\right)$

4. **Write the general statements that give all the solutions.**

   $x = \dfrac{\pi}{6} + \pi n$

   $x = \dfrac{\pi}{6}, \dfrac{7\pi}{6}, \dfrac{13\pi}{6}, \dfrac{19\pi}{6} \cdots$

   These statements mean that all the angles you find by adding or subtracting multiples of π will provide solutions for this equation.

Factoring In the Solutions

The same type of factoring that algebra uses is a great help in solving trig equations. The only trick is to recognize that instead of just x's, y's, or other single-letter variables, trig variables such as $\sin x$ or $\sec y$ exist. Look for the patterns and apply the factoring techniques. Here's a list of the basic factoring patterns.

Factoring binomials:

- ✔ Greatest common factor: $ab \pm cb = b(a \pm c)$
- ✔ Difference of squares: $a^2 - b^2 = (a + b)(a - b)$
- ✔ Sum or difference of cubes: $a^3 + b^3 = (a + b)(a^2 - ab + b^2)$
  $a^3 - b^3 = (a - b)(a^2 + ab + b^2)$

Factoring trinomials:

- Greatest common factor: $ax^2 + ax + ac = a(x^2 + x + c)$
- Un-FOIL: $abx^2 + (ad + bc)x + cd = (ax + c)(bx + d)$

Factoring by grouping: $$\begin{aligned} &abxy + adx + bcy + cd \\ &= ax(by + d) + c(by + d) = (ax + c)(by + d) \end{aligned}$$

Finding a greatest common factor

The trig equations that require finding a greatest common factor, factoring it out, and then solving the equation could look like these two equations: $2\sin x \cos x - \sin x = 0$ or $\cos x \tan x = \sqrt{3}\cos x$. I solve both of these equations in this section.

Solve $2\sin x \cos x - \sin x = 0$ for all the values of x such that $0° \leq x < 360°$.

1. **Factor out sin x from each of the two terms.**

   $\sin x (2\cos x - 1) = 0$

2. **Set the two different factors equal to 0.**

   $\sin x = 0$ or $2\cos x - 1 = 0$

3. **Solve for the values of x that satisfy both equations. Use the table in the Appendix.**

   If $\sin x = 0$, then $x = \sin^{-1}(0) = 0°, 180°$.

   If $2\cos x - 1 = 0$, then $x = \cos^{-1}\left(\dfrac{1}{2}\right) = 60°, 300°$.

   All these values are solutions for the original equation. The complete list is $x = 0°, 60°, 180°, 300°$.

Now solve $\cos x \tan x = \sqrt{3}\cos x$ for all the possible values in degrees.

1. **Move the term on the right to the left by subtracting it from each side.**

   $\cos x \tan x - \sqrt{3}\cos x = 0$

2. **Factor out the cos x from each term.**

   $\cos x \left(\tan x - \sqrt{3}\right) = 0$

   You don't want to divide each side by cos x, because you'll lose a solution if you do.

3. **Set the two different factors equal to 0.**

   $\cos x = 0$ or $\tan x - \sqrt{3} = 0$

4. **Solve for the values of x that satisfy both equations.**

   If $\cos x = 0$, then $x = \cos^{-1}(0) = 90°, 270°$ or $90° + 180°n$.

   If $\tan x - \sqrt{3} = 0$, then $x = \tan^{-1}(\sqrt{3}) = 60°, 240°$ or $60° + 180°n$.

   So the solutions are all of the form $x = 90° + 180°n$ or $x = 60° + 180°n$.

Factoring quadratics

Quadratic equations are nice to work with because, when they don't factor, you can solve them by using the quadratic formula (see the respective section later in this chapter). The types of quadratic trig equations that you can factor are those like $\tan^2 x = \tan x$, $4\cos^2 x - 3 = 0$, $2\sin^2 x + 5\sin x - 3 = 0$, and $\csc^2 x + \csc x - 2 = 0$. Notice that they all have the telltale trig function raised to the second degree. I show you how to handle them in the following examples.

The first two examples have just two terms. One has two variable terms, and the other has just one variable term. In the first example, you put both terms on the left and then factor out the variable or trig term.

Solve $\tan^2 x = \tan x$ for the values of x such that $0 \le x < 2\pi$.

1. **Move the term on the right to the left by subtracting it from both sides.**

   $\tan^2 x - \tan x = 0$

   Don't divide through by $\tan x$. You'll lose solutions.

2. **Factor out $\tan x$.**

   $\tan x (\tan x - 1) = 0$

3. **Set each of the two factors equal to 0.**

   $\tan x = 0$ or $\tan x - 1 = 0$

4. **Solve for the values of x that satisfy both equations.**

   If $\tan x = 0$, then $x = \tan^{-1}(0) = 0, \pi$.

   If $\tan x - 1 = 0$, then $x = \tan^{-1}(1) = \dfrac{\pi}{4}, \dfrac{5\pi}{4}$.

In this next example, the binomial doesn't factor easily as the difference of two squares, because the 3 isn't a perfect square, and you have to use a radical in the factorization. A nice, efficient way to solve this equation is to move the 3 to the right and take the square root of each side.

Solve for all the possible solutions of $4\cos^2 x - 3 = 0$ in degrees.

1. **Move the number to the right by adding 3 to each side.**

$4\cos^2 x = 3$

2. **Take the square root of each side. Then solve for cos x by dividing each side by 2.**

$$\sqrt{4\cos^2 x} = \pm\sqrt{3}$$
$$2\cos x = \pm\sqrt{3}$$
$$\cos x = \pm\frac{\sqrt{3}}{2}$$

3. **Solve the two equations for the values of x.**

If $\cos x = \dfrac{\sqrt{3}}{2}$, then $x = \cos^{-1}\left(\dfrac{\sqrt{3}}{2}\right) = 30°, 330°$

$$x = 30° + 360°n \text{ or } 330° + 360°n$$

If $\cos x = -\dfrac{\sqrt{3}}{2}$, then $x = \cos^{-1}\left(-\dfrac{\sqrt{3}}{2}\right) = 150°, 210°$

$$x = 150° + 360°n \text{ or } 210° + 360°n$$

When you consider all the multiples of 360 degrees added to the four base angles, you find that this equation has lots and lots of solutions.

The next two examples involve using un-FOIL — a technique for determining which two binomials give you a particular quadratic trinomial. Sometimes, when the pattern in the trinomial is obscured, you may want to first substitute some other variable for the trig function to help figure out how you factor it. I do this when I solve $2\sin^2 x + 5\sin x - 3 = 0$ for all the values of x between 0 and 360 degrees.

1. **Replace each sin x with y.**

$2y^2 + 5y - 3 = 0$

2. **Factor the trinomial as the product of two binomials.**

$(2y - 1)(y + 3) = 0$

3. **Replace each y with sin x.**

$(2\sin x - 1)(\sin x + 3) = 0$

4. **Set each factor equal to 0.**

$2\sin x - 1 = 0 \text{ or } \sin x + 3 = 0$

5. **Solve the two equations for the values of x that satisfy them.**

If $2\sin x - 1 = 0$, $2\sin x = 1$, $\sin x = \dfrac{1}{2}$, then $x = \sin^{-1}\left(\dfrac{1}{2}\right) = 30°, 150°$.

If $\sin x + 3 = 0$, $\sin x = -3$, then $x = \sin^{-1}(-3)$. This result is nonsense, because the sine function only produces values between –1 and 1 — so this factor doesn't produce any solutions. Go to Chapter 8 for information on the range of the sine function.

The only two solutions are 30 and 150 degrees.

This next example factors fairly easily, but it involves a reciprocal function. Solve $\csc^2 x + \csc x - 2 = 0$ for any angles between 0 and 2π radians.

1. **Factor the quadratic trinomial into the product of two binomials.**

   $(\csc x + 2)(\csc x - 1) = 0$

2. **Set each factor equal to 0.**

   $\csc x + 2 = 0$ or $\csc x - 1 = 0$

3. **Solve the two equations for the values of x that satisfy them.**

   If $\csc x + 2 = 0$, $\csc x = -2$, then $x = \csc^{-1}(-2) = \dfrac{7\pi}{6}, \dfrac{11\pi}{6}$.

   If $\csc x - 1 = 0$, $\csc x = 1$, then $x = \csc^{-1}(1) = \dfrac{\pi}{2}$.

   An alternate way of dealing with these two binomial equations is to change them by using the reciprocal identity and writing the reciprocal of the number. For the first equation, you'd change from cosecant to sine: $\csc x + 2 = 0$, $\csc x = -2$, $\sin x = -\dfrac{1}{2}$. Do the same for the second equation: $\csc x - 1 = 0$, $\csc x = 1$, $\sin x = 1$. You'd then solve the inverse equations.

Increasing the degrees in factoring

Factoring quadratics is a breeze — well, I guess it gets a bit windy at times. Factoring equations with higher degrees can get a bit nasty if you don't have a nice situation such as just two terms or a quadraticlike equation. You may have the possibility of factoring by grouping, but I cover that method in the next section. In this section, the problems that I have in mind are those like $2\sin^3 x = \sin x$ and $2\cos^4 x - 9\cos^2 x + 4 = 0$.

The first equation has just two terms, so you can factor it by finding a greatest common factor. Solve $2\sin^3 x = \sin x$ for all the possible angles in degrees.

1. **Move the term on the right to the left by subtracting it from each side.**

   $2\sin^3 x - \sin x = 0$

2. **Factor out sin x.**

   $\sin x \left(2\sin^2 x - 1 = 0\right)$

3. **Set each factor equal to 0.**

   $\sin x = 0$ or $2\sin^2 x - 1 = 0$

4. **Solve the two equations for the values of x that satisfy them.**

   If $\sin x = 0$, then $x = \sin^{-1}(0) = 0°, 180°, \dots$ or $0° + 180°n$.

   If $2\sin^2 x - 1 = 0$, $2\sin^2 x = 1$, $\sin^2 x = \frac{1}{2}$, then you end up with a quadratic equation.

5. **Take the square root of both sides of the equation and solve for x.**

   Multiply both parts of the fraction by the denominator to get the radical out of the denominator.

   $$\sqrt{\sin^2 x} = \pm \sqrt{\frac{1}{2}}$$

   $$\sin x = \pm \frac{\sqrt{2}}{2}$$

   Now, considering both solutions:

   If $\sin x = \frac{\sqrt{2}}{2}$, then $x = \sin^{-1}\left(-\frac{\sqrt{2}}{2}\right) = 45°, 135°, \dots$

   or $45° + 360°n, \ 135° + 360°n$

   If $\sin x = -\frac{\sqrt{2}}{2}$, then $x = \sin^{-1}\left(-\frac{\sqrt{2}}{2}\right) = 225°, 315°, \dots$

   or $225° + 360°n, \ 315° + 360°n$

This fourth-degree trig equation has a whole slew of answers:

$x = 180°n$

$x = 45° + 360°n$

$x = 135° + 360°n$

$x = 225° + 360°n$

$x = 315° + 360°n$

You can combine these last four equations for x, the ones that begin with multiples of 45 degrees, to read $x = 45° + 90°n$. This equation generates all the same angles as the last four statements combined. How do you know you can simplify this way? Because the angles of 45, 135, 225, and 315 degrees are all 90 degrees apart in value. By starting with the 45 and adding 90 over and over, you get all the listed angles as well as the infinite number of their multiples.

The next example is also a fourth-degree equation, but this one is quadratic-like, meaning that it factors like a quadratic trinomial with two binomial factors. This problem has the possibility of having a great number of solutions — or none. Solve $2\cos^4 x - 9\cos^2 x + 4 = 0$ for the solutions that are between 0 and 2π.

1. **Factor the trinomial as the product of two binomials.**

   $(2\cos^2 x - 1)(\cos^2 x - 4) = 0$

2. **Set each factor equal to 0.**

   $2\cos^2 x - 1 = 0$ or $\cos^2 x - 4 = 0$

3. **Solve for the function in each equation by getting the terms with cosine in them alone on one side of the equation.**

   $2\cos^2 x - 1 = 0,\ 2\cos^2 x = 1,\ \cos^2 x = \dfrac{1}{2}$

   $\cos^2 x - 4 = 0,\ \cos^2 x = 4$

4. **Take the square root of each side of each equation.**

   $$\sqrt{\cos^2 x} = \pm\sqrt{\dfrac{1}{2}}$$

   $$\cos x = \pm\dfrac{\sqrt{2}}{2}$$

   $$\sqrt{\cos^2 x} = \pm\sqrt{4}$$

   $$\cos x = \pm 2$$

5. **Solve for the values of x that satisfy the equations.**

   If $\cos x = \dfrac{\sqrt{2}}{2}$, then $x = \cos^{-1}\left(\dfrac{\sqrt{2}}{2}\right) = \dfrac{\pi}{4}, \dfrac{7\pi}{4}$.

   If $\cos x = -\dfrac{\sqrt{2}}{2}$, then $x = \cos^{-1}\left(-\dfrac{\sqrt{2}}{2}\right) = \dfrac{3\pi}{4}, \dfrac{5\pi}{4}$.

   If $\cos x = \pm 2$, then you have a problem — that equation doesn't compute! The cosine function results in values between –1 and 1. (Find out more about the range of cosine in Chapter 8.) This factor doesn't give any new solutions to the original problem.

Factoring by grouping

The process of factoring by grouping works in very special cases, when the original equation is the result of multiplying two binomials together that have some unrelated terms in them. You usually can apply this type of factoring when you're facing an even number of terms and can find common factors in different groups of them. The types of equations that you can solve by using grouping look like $4\sin x\cos x - 2\sin x - 2\cos x + 1 = 0$ or $\sin^2 x\sec x + 2\sin^2 x = \sec x + 2$. In the first equation, the first two terms have an obvious common factor, $2\sin^2 x$. The second two have no common factor other than 1, but to make grouping work, factor out –1.

Solve $4\sin x\cos x - 2\sin x - 2\cos x + 1 = 0$ for all the possible answers between 0 and 2π.

1. **Factor $2\sin x$ out of the first two terms and –1 out of the second two.**

   $2\sin x(2\cos x - 1) - 1(2\cos x - 1) = 0$

   Now you have two terms, each with a factor of $2\cos x - 1$.

2. **Factor that common factor out of the two terms.**

   $(2\cos x - 1)(2\sin x - 1) = 0$

3. **Set the two factors equal to 0.**

   $2\cos x - 1 = 0,\ 2\cos x = 1,\ \cos x = \dfrac{1}{2}$

   $2\sin x - 1 = 0,\ 2\sin x = 1,\ \sin x = \dfrac{1}{2}$

4. **Solve for the values of x that satisfy the equation.**

   If $\cos x = \dfrac{1}{2}$, then $x = \cos^{-1}\left(\dfrac{1}{2}\right) = \dfrac{\pi}{3}, \dfrac{5\pi}{3}$.

   If $\sin x = \dfrac{1}{2}$, then $x = \sin^{-1}\left(\dfrac{1}{2}\right) = \dfrac{\pi}{6}, \dfrac{5\pi}{6}$.

This next example of grouping requires that you begin by moving the two terms on the right to the left. Another twist is that one of the resulting factors turns out to be a quadratic. How can math be much more fun than this?

Solve $\sin^2 x\sec x + 2\sin^2 x = \sec x + 2$ for all the angles between 0 and 360 degrees.

1. **Move the terms on the right to the left by subtracting them from both sides.**

   $\sin^2 x\sec x + 2\sin^2 x - \sec x - 2 = 0$

2. **Factor $\sin^2 x$ out of the first two terms and –1 out of the second two.**

   $\sin^2 x(\sec x + 2) - 1(\sec x + 2) = 0$

3. **Now factor $\sec x + 2$ out of the two terms.**

   $(\sec x + 2)(\sin^2 x - 1) = 0$

4. **Set the two factors equal to 0.**

   $\sec x + 2 = 0$, $\sec x = -2$

   $\sin^2 x - 1 = 0$, $\sin^2 x = 1$, $\sin x = \pm 1$ when you take the square root of both sides.

5. **Solve for the values of x that satisfy the equations.**

   If $\sec x = -2$, then $x = \sec^{-1}(-2) = 120°, 240°$.

   If $\sin x = 1$, then $x = \sin^{-1}(1) = 90°$.

   If $\sin x = -1$, then $x = \sin^{-1}(-1) = 270°$.

Using the Quadratic Formula

When quadratic equations factor, life is good. When they don't, you can still survive, thanks to that wonderful quadratic formula. In case you've forgotten the exact formula, here it is.

The quadratic formula says that if you have a quadratic equation in the form $ax^2 + bx + c = 0$, then its solutions are $x = \dfrac{-b \pm \sqrt{b^2 - 4ac}}{2a}$.

In trig, a trig function replaces the x or variable part of the quadratic formula. For example, find the solution of $\sin^2 x - 4\sin x - 1 = 0$ for all angles between 0 and 360 degrees. Instead of just x's, the variable terms are $\sin x$'s.

1. **Identify the values of the a, b, and c in the formula.**

   The values are $a = 1$, $b = -4$, and $c = -1$.

2. **Fill in the quadratic formula with these values and simplify.**

   $$\sin x = \frac{-(-4) \pm \sqrt{(-4)^2 - 4(1)(-1)}}{2(1)}$$

   $$= \frac{4 \pm \sqrt{16+4}}{2} = \frac{4 \pm \sqrt{20}}{2}$$

   $$= \frac{4 \pm 2\sqrt{5}}{2} = 2 \pm \sqrt{5}$$

3. **Find approximate values for $\sin x$ from the solved form.**

   Using a scientific calculator, $2 \pm \sqrt{5} \approx 2 \pm 2.236$. So $\sin x$ is either about 4.236 or –0.236.

4. **Use a table of values to find approximate angles with these sines.**

   If $\sin x = 4.236$, you get an impossible result. The value of the sine ranges from –1 to 1, so $\sin x$ can't have this value.

If $\sin x = -0.236$, then $x = \sin^{-1}(-0.236) \approx -14°$ or $346°$. These are the same angle. First, you write it as a negative angle and then as its positive equivalent.

Another angle satisfies this equation, too. The other negative angle that has a 14-degree reference angle is the third-quadrant angle of 194 degrees. Refer to Chapter 8 for more on reference angles.

Incorporating Identities

Some trig equations contain more than one trig function. Others have mixtures of multiple angles and single angles with the same variable. Some examples of such equations include $3\cos^2 x = \sin^2 x$, $2\sec x = \tan x + \cot x$, $\cos 2x + \cos x + 1 = 0$, and $\sin x \cos x = \frac{1}{2}$. To get these equations into a more-manageable form so that you can use factoring or one of the other methods in this chapter to solve them, you use identities to substitute some or all of the terms (for more on trig identities, see Chapter 10). For example, to solve $3\cos^2 x = \sin^2 x$ for all the angles between 0 and 2π, apply the Pythagorean identity.

1. **Replace $\sin^2 x$ with its equivalent from the Pythagorean identity, $\sin^2 x + \cos^2 x = 1$ or $\sin^2 x = 1 - \cos^2 x$.**

   $3\cos^2 x = 1 - \cos^2 x$

2. **Add $\cos^2 x$ to each side and simplify by dividing.**

   $4\cos^2 x = 1$

   $\cos^2 x = \frac{1}{4}$

3. **Take the square root of each side.**

   $\sqrt{\cos^2 x} = \pm\sqrt{\frac{1}{4}}$

   $\cos x = \pm\frac{1}{2}$

4. **Solve for the values of x that satisfy the equation.**

   If $\cos x = \frac{1}{2}$, then $x = \cos^{-1}\left(\frac{1}{2}\right) = \frac{\pi}{3}, \frac{5\pi}{3}$.

   If $\cos x = -\frac{1}{2}$, then $x = \cos^{-1}\left(-\frac{1}{2}\right) = \frac{2\pi}{3}, \frac{4\pi}{3}$.

In this next example, you begin with three different trig functions. A good tactic is to replace each function by using either a ratio identity or a reciprocal identity. Using these identities creates fractions, and fractions require common denominators. By the way, having fractions in trig equations is *good,* because the products that result from multiplying and making equivalent fractions are usually parts of identities that you can then substitute in to make the expression much simpler. Solve $2\sec x = \tan x + \cot x$ for all the possible solutions in degrees.

1. **Replace each term with its respective reciprocal or ratio identity.**

$$2\left(\frac{1}{\cos x}\right) = \frac{\sin x}{\cos x} + \frac{\cos x}{\sin x}$$

$$\frac{2}{\cos x} = \frac{\sin x}{\cos x} + \frac{\cos x}{\sin x}$$

2. **Rewrite the fractions with the common denominator $\sin x \cos x$.**

   Multiply each term by a fraction that equals 1, with either sine or cosine in both the numerator and denominator.

$$\frac{2}{\cos x} \cdot \frac{\sin x}{\sin x} = \frac{\sin x}{\cos x} \cdot \frac{\sin x}{\sin x} + \frac{\cos x}{\sin x} \cdot \frac{\cos x}{\cos x}$$

$$\frac{2\sin x}{\sin x \cos x} = \frac{\sin^2 x}{\sin x \cos x} + \frac{\cos^2 x}{\sin x \cos x}$$

3. **Add the two fractions on the right. Then, using the Pythagorean identity, replace the new numerator with 1.**

$$\frac{2\sin x}{\sin x \cos x} = \frac{\sin^2 x + \cos^2 x}{\sin x \cos x}$$

$$\frac{2\sin x}{\sin x \cos x} = \frac{1}{\sin x \cos x}$$

4. **Set the equation equal to 0 by subtracting the right term from each side.**

$$\frac{2\sin x - 1}{\sin x \cos x} = 0$$

5. **Now set the numerator equal to 0.**

$$2\sin x - 1 = 0 \quad \text{or} \quad 2\sin x = 1, \ \sin x = \frac{1}{2}$$

   If the numerator is equal to 0, then the whole fraction is equal to 0. The denominator can't equal 0 — such a number doesn't exist.

6. **Solve for the values of x that satisfy the original equation.**

$$x = \sin^{-1}\left(\frac{1}{2}\right) = 30°, 150°, \ldots$$

$$x = 30° + 360°n \quad \text{or} \quad x = 150° + 360°n$$

In the next example, two different angles are in play. One angle is twice the size of the other, so you use a double-angle identity to reduce the terms to only one angle. The trick is to choose the correct version of the cosine double-angle identity.

Solve $\cos 2x + \cos x + 1 = 0$ for x between 0 and 2π.

1. Replace $\cos 2x$ with $2\cos^2 x - 1$.

$2\cos^2 x - 1 + \cos x + 1 = 0$

This version of the cosine double-angle identity is preferable because the other trig function in the equation already has a cosine in it.

2. Simplify the equation. Then factor out $\cos x$.

$2\cos^2 x + \cos x = 0$

$\cos x (2\cos x + 1) = 0$

3. Set each factor equal to 0.

$\cos x = 0$ or $2\cos x + 1 = 0$, $2\cos x = -1$, $\cos x = -\dfrac{1}{2}$

4. Solve for the values of x that satisfy the original equation.

If $\cos x = 0$, then $x = \cos^{-1}(0) = \dfrac{\pi}{2}, \dfrac{3\pi}{2}$.

If $\cos x = -\dfrac{1}{2}$, then $x = \cos^{-1}\left(-\dfrac{1}{2}\right) = \dfrac{2\pi}{3}, \dfrac{4\pi}{3}$.

This last example can be deceptively simple. The catch is that you have to recognize a double-angle identity upfront and make a quick switch. This example is also a nice segue into the next section on equations with multiple–angle solutions. Solve $\sin x \cos x = \dfrac{1}{2}$ for all the solutions between 0 and 360 degrees.

1. Use the sine double-angle identity to create a substitution for the expression on the left.

Starting out with the identity and multiplying each side by $\dfrac{1}{2}$, you get

$\sin 2x = 2\sin x \cos x$

$\dfrac{1}{2}\sin 2x = \sin x \cos x$

2. Replace the expression on the left of the original equation with its equivalent from the double-angle identity.

$\sin x \cos x = \dfrac{1}{2}$

$\dfrac{1}{2}\sin 2x = \dfrac{1}{2}$

3. Multiply each side of the equation by 2.

$2 \cdot \dfrac{1}{2}\sin 2x = \dfrac{1}{2} \cdot 2$

$\sin 2x = 1$

4. Rewrite the expression as an inverse function.

$$2x = \sin^{-1}(1)$$

See Chapter 13 for more on inverse functions.

5. Determine which angles within *two* rotations satisfy the expression.

$$2x = \sin^{-1}(1) = 90°, 450°$$

You use two rotations because the coefficient of x is 2.

6. Divide each term by 2.

$$\frac{2x}{2} = \frac{90°}{2}, \frac{450°}{2}$$
$$x = 45°, 225°$$

Notice that the resulting angles are between 0 and 360 degrees.

You can generalize the double-angle technique from the preceding example for other multiple-angle expressions.

Finding Multiple-Angle Solutions

Multiple-angle expressions are those where the angle measure is some multiple of a variable — for example, $2x$ or $3y$. In this section, I show you how to take these expressions apart and solve for all the additional solutions that are possible. Because the trig functions are *periodic* (meaning they repeat their patterns infinitely), the number of possibilities for solutions increases tremendously. The larger the multiplier, the more solutions.

When solving a trig equation of the form $ax = f^{-1}(x)$ where you want the solution to be all the angles within one complete rotation, write out all the solutions within the number of complete rotations that a represents. Then divide each angle measure by a.

Problems that lend themselves to this technique are those such as $2\sin^2 5x = 1$ and $\cos\left(\frac{1}{2}x\right) = -\frac{\sqrt{3}}{2}$. In the first example, I solve $2\sin^2 5x = 1$ for all the angles between 0 and 2π.

1. Divide each side by 2; then take the square root of each side.

$$\sin^2 5x = \frac{1}{2}$$

$$\sqrt{\sin^2 5x} = \pm\sqrt{\frac{1}{2}} = \pm\frac{\sqrt{2}}{2}$$

$$\sin 5x = \pm\frac{\sqrt{2}}{2}$$

2. **Solve for 5x, which represents the angles that satisfy the equation within one rotation.**

   If $\sin 5x = \frac{\sqrt{2}}{2}$, then $5x = \sin^{-1}\left(\frac{\sqrt{2}}{2}\right) = \frac{\pi}{4}, \frac{3\pi}{4}$.

   If $\sin 5x = -\frac{\sqrt{2}}{2}$, then $5x = \sin^{-1}\left(-\frac{\sqrt{2}}{2}\right) = \frac{5\pi}{4}, \frac{7\pi}{4}$.

3. **Extend the solutions to five rotations by adding 2π to each of the original angles four times.**

   $$5x = \frac{\pi}{4}, \frac{3\pi}{4}, \frac{9\pi}{4}, \frac{11\pi}{4}, \frac{17\pi}{4}, \frac{19\pi}{4}, \frac{25\pi}{4}, \frac{27\pi}{4}, \frac{33\pi}{4}, \frac{35\pi}{4}$$

   $$5x = \frac{5\pi}{4}, \frac{7\pi}{4}, \frac{13\pi}{4}, \frac{15\pi}{4}, \frac{21\pi}{4}, \frac{23\pi}{4}, \frac{29\pi}{4}, \frac{31\pi}{4}, \frac{37\pi}{4}, \frac{39\pi}{4}$$

4. **Divide all the terms by 5 and simplify.**

   $$x = \frac{\pi}{20}, \frac{3\pi}{20}, \frac{9\pi}{20}, \frac{11\pi}{20}, \frac{17\pi}{20}, \frac{19\pi}{20}, \frac{25\pi}{20}, \frac{27\pi}{20}, \frac{33\pi}{20}, \frac{35\pi}{20}$$
   $$x = \frac{\pi}{20}, \frac{3\pi}{20}, \frac{9\pi}{20}, \frac{11\pi}{20}, \frac{17\pi}{20}, \frac{19\pi}{20}, \frac{5\pi}{4}, \frac{27\pi}{20}, \frac{33\pi}{20}, \frac{7\pi}{4}$$
   $$x = \frac{5\pi}{20}, \frac{7\pi}{20}, \frac{13\pi}{20}, \frac{15\pi}{20}, \frac{21\pi}{20}, \frac{23\pi}{20}, \frac{29\pi}{20}, \frac{31\pi}{20}, \frac{37\pi}{20}, \frac{39\pi}{20}$$
   $$x = \frac{\pi}{4}, \frac{7\pi}{20}, \frac{13\pi}{20}, \frac{3\pi}{4}, \frac{21\pi}{20}, \frac{23\pi}{20}, \frac{29\pi}{20}, \frac{31\pi}{20}, \frac{37\pi}{20}, \frac{39\pi}{20}$$

   Notice that all 16 solutions are angles with measures less than 2π.

This next example has a proper-fraction multiplier rather than a multiplier greater than 1.

Solve $\cos\left(\frac{1}{2}x\right) = -\frac{\sqrt{3}}{2}$ for all the solutions between 0 and 360 degrees.

1. **Rewrite the equation as an inverse trig equation.**

   $$\frac{1}{2}x = \cos^{-1}\left(-\frac{\sqrt{3}}{2}\right)$$

2. **Determine which angles satisfy the inverse equation within one full rotation.**

   $$\frac{1}{2}x = \cos^{-1}\left(-\frac{\sqrt{3}}{2}\right) = 150°, 210°$$

3. **Multiply all the terms by 2.**

   $$x = 300°, 420°$$

4. **Throw out the second angle, because its measure is greater than 360 degrees.**

   The only solution is 300 degrees. When you replace the x in the original equation with this angle measure, you get a true statement.

Squaring Both Sides

When solving trig equations, you have so many choices to choose among. Many times, more than one method will work — although one method is usually quicker or easier than another. And then you'll come across a trig equation that defies your finest attempts. Two last-ditch efforts that you can use when solving trig equations are to square both sides of the equation or to multiply each term through by a trig function that you've carefully selected. I show you the first of these two methods here, and I show you the second method in the next section. Examples of equations that respond well to squaring both sides include $\sin x + \cos x = \sqrt{2}$ and $\cos x - \sqrt{3}\sin x = 1$.

Solve $\sin x + \cos x = \sqrt{2}$ for all the possible angles in degrees.

1. Square both sides of the equation.

When squaring a binomial, be sure not to forget the middle term.

$$(\sin x + \cos x)^2 = (\sqrt{2})^2$$
$$\sin^2 x + 2\sin x \cos x + \cos^2 x = 2$$

2. Use the Pythagorean identity to replace $\sin^2 x + \cos^2 x$.

$$\sin^2 x + \cos^2 x + 2\sin x \cos x = 2$$
$$1 + 2\sin x \cos x = 2$$

3. Subtract 1 from each side. Then replace the expression on the left with the sine double-angle formula.

$$2\sin x \cos x = 1$$
$$\sin 2x = 1$$

4. Solve for the value of $2x$ by using the inverse function. Then write a few angle solutions to determine a pattern.

$$2x = \sin^{-1}(1) = 90°, 450°, 810°\ldots$$

Because you're supposed to find all the possible solutions, you're not bound by only two rotations.

5. Divide every term by 2.

$$2x = 90°, 450°, 810°\ldots$$
$$x = 45°, 225°, 405°\ldots$$

6. Write an expression for all the solutions.

$$x = 45° + 180°n$$

In the next example, you need to do a little shifting at first. To solve $\cos x - \sqrt{3}\sin x = 1$, get the term with the radical in it to one side of the equation by itself. Otherwise, when you square both sides, you end up with a radical factor in one of the terms. That situation isn't always bad, but dealing with it is usually a little more awkward than not.

Solve the equation for all the possible angles from 0 to 360 degrees.

1. **Add the radical term to both sides and subtract 1 from both sides.**

   You get $\cos x - 1 = \sqrt{3} \sin x$.

2. **Square both sides.**

   $$(\cos x - 1)^2 = (\sqrt{3} \sin x)^2$$
   $$\cos^2 x - 2\cos x + 1 = 3\sin^2 x$$

3. **Replace $\sin^2 x$ with $1 - \cos^2 x$ from the Pythagorean identity.**

   Doing so creates an equation with terms that have all the same functions, $\cos x$, in them.

   $$\cos^2 x - 2\cos x + 1 = 3(1 - \cos^2 x)$$

4. **Simplify the equation by distributing the 3 on the right and then bringing all the terms to the left to set the equation equal to 0.**

   $$\cos^2 x - 2\cos x + 1 = 3 - 3\cos^2 x$$
   $$4\cos^2 x - 2\cos x - 2 = 0$$

5. **Divide every term by 2.**

   $$2\cos^2 x - \cos x - 1 = 0$$

6. **Factor the quadratic equation.**

   $$(2\cos x + 1)(\cos x - 1) = 0$$

7. **Set each factor equal to 0.**

   $$2\cos x + 1 = 0 \quad \text{or} \quad \cos x - 1 = 0$$
   $$2\cos x = -1 \qquad\qquad \cos x = 1$$
   $$\cos x = -\frac{1}{2}$$

8. **Solve each equation for the value of x.**

   If $\cos x = -\frac{1}{2}$, then $x = \cos^{-1}\left(-\frac{1}{2}\right) = 240°, 300°$.

   If $\cos x = 1$, then $x = \cos^{-1}(1) = 0°, 360°$.

   The angles 0 and 360 degrees have the same terminal side. You usually list just one of them: 0 degrees.

Multiplying Through

The technique of multiplying through a trig equation by a carefully selected function shouldn't be your first choice — or your second, third, or fourth choice. This method is usually a last resort. Not that the method is terribly hard; it just requires sitting back and looking at the equation, and magically coming up with the best function to multiply through by. You can find the best function by guess or by golly, but then, that would take all the fun out of it — you want to guess right the first time. Here's an example of an equation that this technique works well on.

Solve $2 \sin x - \csc x = 1$ for all the solutions between 0 and 2π.

1. **Multiply each term by sin x.**

   Why sin x? I chose that function because I could see that the products of the individual terms would be either different powers of sine or just a number. Notice that the product of csc x and its reciprocal, sin x, is 1.

   $$2 \sin x \cdot \sin x - \csc x \cdot \sin x = 1 \cdot \sin x$$
   $$2 \sin^2 x - 1 = \sin x$$

2. **Subtract sin x from each side to set the equation equal to 0.**

   $$2 \sin^2 x - \sin x - 1 = 0$$

3. **Factor the quadratic equation.**

   $$(2 \sin x + 1)(\sin x - 1) = 0$$

4. **Set each factor equal to 0.**

   $$2 \sin x + 1 = 0 \qquad \text{or} \qquad \sin x - 1 = 0$$
   $$\sin x = -\frac{1}{2} \qquad\qquad \sin x = 1$$

5. **Solve for the values that satisfy the equations.**

   If $\sin x = -\frac{1}{2}$, then $x = \sin^{-1}\left(-\frac{1}{2}\right) = \frac{7\pi}{6}, \frac{11\pi}{6}$.

   If $\sin x = 1$, then $x = \sin^{-1}(1) = \frac{\pi}{2}$.

Solving with a Graphing Calculator

Some of the more-advanced graphing calculators can make short work of solving trig equations. A graphing calculator comes in very handy when the equation is complicated, has several different functions or angle multiples, or has fractional or decimal values that don't lend themselves to the traditional solving methods that I discuss throughout this chapter. For example, I prefer to use a graphing calculator to solve equations like $\cos 2x = 2 \cos x$ and $\cos^2 x - 0.4 \sin x = 0.6$.

First, here's how to solve $\cos 2x = 2 \cos x$ for all solutions between -2π and 2π.

1. **Put the expression on the left side of the equation in the y menu (the graphing menu) of your calculator. Put the expression on the right as a second entry.**

   $y_1 = \cos 2x$

   $y_2 = 2 \cos x$

2. **Set the window of your calculator to show the graphs.**

   Set the x values from -2π to 2π. In decimal form, let $x = -6.5$ to 6.5 to give a little room on either side of the left and right ends.

   Set the y values to go from -3 to 3. Doing so gives room above and below the graph. If you have an auto fit capability, use it to make the graph fit automatically after you choose the x values you want the graph to encompass.

3. **Graph the two functions, and see where they intersect (see Figure 14-1).**

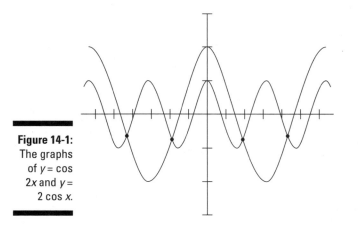

Figure 14-1: The graphs of $y = \cos 2x$ and $y = 2 \cos x$.

4. **Use the *intersect* feature on the calculator to determine the solutions.**

   The x-coordinates of the intersection points are the solutions (rounded to four decimal places): $x = -4.3377$, -1.9455, 1.9455, and 4.3377. These solutions are in radians — the π value is already multiplied through.

You can also find the solutions to the preceding example with a graphing calculator's *solver* feature, but usually, you still need to look at the graph anyway so you know how many solutions you're trying to find. The solver feature usually finds only one solution at a time, and you need a hint to know where to find them.

This next example has decimals built in, so you probably can't factor it. You can solve it by using identities and writing it as a quadratic, and then using the quadratic formula. This calculator method gives you another option.

Solve $\cos^2 x - 0.4 \sin x = 0.6$ for all angles between $-\pi$ and π.

1. **Put the expression on the left in the graphing y menu of your calculator. Put the expression on the right as a second entry.**

   $y_1 = \cos^2 x - 0.4 \sin x$

   $y_2 = 0.6$

2. **Set the window of your calculator to show the graphs.**

   Set the horizontal, x values from $-\pi$ to π. In decimal form, use $x = -3.2$ to 3.2 to give a little room on either side of the ends.

   Set the vertical, y values to go from -3 to 3. Doing so gives room above and below the graph. If you have an auto fit capability, use it to make the graph fit automatically.

3. **Graph the two functions, and see where they intersect (see Figure 14-2).**

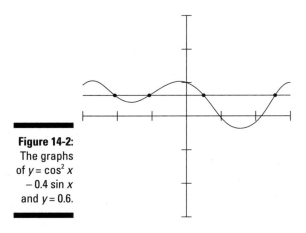

Figure 14-2:
The graphs
of $y = \cos^2 x$
$- 0.4 \sin x$
and $y = 0.6$.

4. **Use the _intersect_ feature on the calculator to determine the solutions.**

   The x-coordinates of the intersection points are the solutions (rounded to four decimal places): $x = -2.0998, -1.0418, 0.4817,$ and 2.6598. These solutions are in radians — the π value is already multiplied through.

Chapter 15

Obeying the Laws

Triangles are very useful figures. Since humankind figured out how to keep records, people have documented the applications of triangles in mathematics and many other sciences. The right triangle gets the most use; Pythagoras saw to it that others recognized right triangles for the powerful polygon that they are. But *oblique* triangles (those that aren't right triangles) have their place, too. You can't always arrange to have a nice right triangle when you want it. Here's where oblique triangles and the laws of sines and cosines come into play.

The law of sines uses, believe it or not, the sines of a triangle's angles. With three carefully selected parts of the triangle, you can solve for the sizes of the other parts. Of course, you have to obey the law, and the choices you can make are limited. That's where the law of cosines comes in to save the day. This law isn't as user-friendly, but it picks up where the law of sines falls short. And to rescue you when all else fails, the law of tangents throws out a lifejacket (although this law isn't nearly as popular as the other two).

Trigonometry opens up all sorts of possibilities for solving area problems. By using the tools in this chapter, you won't find a triangle that you can't lick.

Describing the Parts of Triangles

When it comes to triangles, you'll find right triangles and nonright triangles, called *oblique* triangles. Other classifications exist too, such as acute, obtuse, isosceles, and so on. But no matter what you call it, any triangle has exactly six parts: three angles and three sides. After you have information about the parts of a triangle, you can make all sorts of computations and manipulations, using triangles to model situations and solve problems.

Standardizing the parts

Usually, when you name the parts of a triangle, you follow a system or pattern. Having this system helps sort out the information, even when you don't have a picture of the triangle to help you. The most common system is to name the angles of the triangle with capital letters, usually A, B, and C, and name the sides opposite each of the angles with the lowercase letter that matches. Figure 15-1 shows what this labeling looks like.

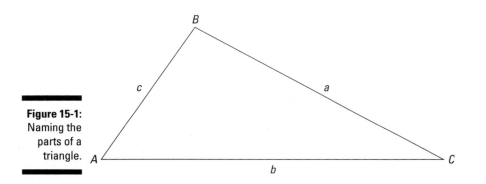

Figure 15-1:
Naming the parts of a triangle.

Another common practice is to name the angles with Greek letters, such as α, β, and γ, but in this chapter, I stick with the capital and lowercase letters.

Determining a triangle

Even though every triangle has six parts, you only need to have information on or know the measures of three particular parts to determine the others. For example, if you know the measures of the three sides, then you know that the three angles are uniquely determined. You can't construct more than one shape and size of triangle from those three sides.

After you know the values for three carefully chosen pieces of a triangle, then you can use any of the three different rules or laws that allow you to find the other three parts of the triangle. I discuss all three laws in their own sections later in this chapter.

Finding the one and only

Several combinations of parts uniquely determine a triangle. Here's a list of them all.

To *uniquely determine* a triangle (find only one possible shape), you need

- ✔ **SSS:** The measures of the three sides
- ✔ **SAS:** The measures of two sides and the angle between them
- ✔ **ASA:** The measures of two angles and the side between them
- ✔ **AAS:** The measures of two angles and one of the sides

The last rule is actually just another version of the one directly before it. When you have two angles, you can determine the third, so the side lies between two known angles. Figure 15-2 shows these situations.

You may have noticed that I didn't mention one combination — AAA, where all three angles are the same. I left it out on purpose because, in such a case, all you can be sure of is that the two triangles are *similar* — they're the same shape but not necessarily the same size.

Dealing with the ambiguous case

Four situations allow you to uniquely determine a triangle, and I list them in the preceding section. One other case can be helpful, even though you may end up with two different triangles instead of one unique triangle: SSA, the measures of two sides and an angle that *isn't* between them. This situation is a little tricky, because often, two different triangles are possible — which is why it's known as the *ambiguous case*. Sometimes, this case is still better than nothing — as long as you're aware that more than one triangle can exist. Figure 15-3 illustrates such a situation. In the two triangles, sides *a* and *c* and angle *A* are the same measure. The angle measure *B* and the length of side *b*, however, aren't the same in both triangles.

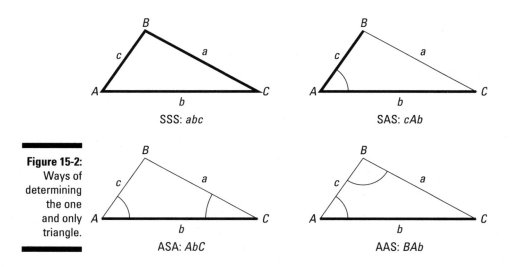

Figure 15-2:
Ways of determining the one and only triangle.

SSS: *abc*

SAS: *cAb*

ASA: *AbC*

AAS: *BAb*

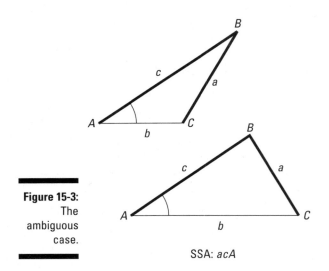

SSA: *acA*

Following the Law of Sines

When you already have two angles, as in the case of ASA or AAS (see the pre-ceding section), you can use the *law of sines.* This law uses the ratios of the sides of a triangle and their opposite angles. The bigger a side, the bigger its opposite angle. The longest side is always opposite the largest angle. Here's how it goes.

The law of sines for triangle *ABC* with sides *a, b,* and *c* opposite those angles, respectively, says

$$\frac{\sin A}{a} = \frac{\sin B}{b} = \frac{\sin C}{c} \text{ and } \frac{a}{\sin A} = \frac{b}{\sin B} = \frac{c}{\sin C}$$

So the law of sines says that in a single triangle, the ratio of each side to its angle is equal to the ratio of any other side to its angle.

For example, consider a triangle where side *a* is 86 inches long and angles *A* and *B* are 84 and 58 degrees, respectively. Figure 15-4 shows a picture of the triangle, and the following steps show you how to find the missing three parts.

1. **Find the measure of angle *C*.**

   The sum of the measures of a triangle's angles is 180 degrees. So find the sum of angles *A* and *B,* and subtract that sum from 180.

   $180 - (84 + 58) = 180 - 142 = 38$

   Angle *C* measures 38 degrees.

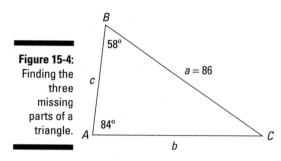

Figure 15-4:
Finding the three missing parts of a triangle.

2. **Find the measure of side *b*.**

• Using the law of sines and the proportion $\frac{a}{\sin A} = \frac{b}{\sin B}$, fill in the values that you know.

$$\frac{86}{\sin 84°} = \frac{b}{\sin 58°}$$

Use the given values, not those that you've determined yourself. That way, if you make an error, you can spot it easier later.

• Use the table in the Appendix or a calculator to determine the values of the sines.

$$\frac{86}{0.995} = \frac{b}{0.848}$$

• Multiply each side by the denominator under *b* to solve for that length. Because the original measures are whole numbers, round this answer to the nearer whole number.

$$(0.848)\frac{86}{0.995} = \frac{b}{0.848}(0.848)$$
$$73.294 = b$$

Side *b* measures about 73 inches.

3. **Find the measure of side *c*.**

• Using the law of sines and the proportion $\frac{a}{\sin A} = \frac{c}{\sin C}$, fill in the values that you know.

$$\frac{86}{\sin 84°} = \frac{c}{\sin 38°}$$

Again, it's best to use the given values, not those that you determined. In this case, however, you have to use a computed value, the angle *C*.

• Use the table in the Appendix or a calculator to determine the values of the sines.

$$\frac{86}{0.995} = \frac{c}{0.616}$$

• Multiply each side by the denominator under c to solve for that length. Because the original measures were given as whole numbers, round this answer to the nearer whole number.

$$(0.616)\frac{86}{0.995} = \frac{c}{0.616}(0.616)$$
$$53.242 = c$$

Side c measures about 53 inches.

Although knowing how to find the missing measures in an oblique triangle seems wonderful, you may wonder, "What's the point?" One major reason for solving triangles is so you can apply them to practical problems. For example, the question, "How tall is it?" seems to be a reasonable request.

Suppose a tree is growing on a hillside. The tree is completely vertical, but the hillside inclines at a 10-degree angle from the horizontal. Josh is standing 100 feet downhill from the tree. The angle of inclination from Josh's feet to the top of the tree is 32 degrees. How tall is the tree? First, take a look at a visual of the situation in Figure 15-5 and then review the steps that follow.

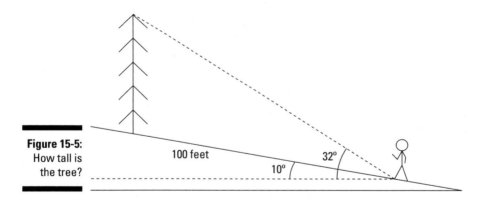

Figure 15-5:
How tall is
the tree?

100 feet

32°

10°

1. **Determine the oblique triangle that you can use to solve the problem.**

   You know that one side is 100 feet long, and you can determine two angles, so use the triangle that Figure 15-6 shows.

2. **Determine the two angles on either side of the base of the triangle.**

   • You determine the angle on the right of the 100-foot base by subtracting the hill's 10-degree inclination from the tree's 32-degree inclination: $32 - 10 = 22$ degrees. (These angles are also known as *angles of elevation,* which you can find out more about in Chapter 9.)

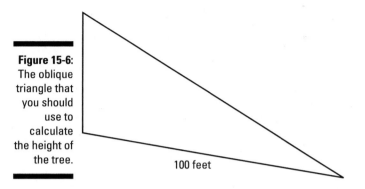

Figure 15-6:
The oblique
triangle that
you should
use to
calculate
the height of
the tree.

100 feet

- The angle on the left of the 100-foot base is supplementary to the
  angle in the right triangle that you can draw below the oblique triangle (see Figure 15-7). Drawing a right angle with the vertical leg following the tree, you determine an angle of 80 degrees by adding the
  90-degree angle and the 10-degree angle and subtracting that sum
  from 180, the total number of degrees in a triangle. *Supplementary*
  angles also add up to 180, so the angle supplementary to the
  80-degree angle is 100 degrees. Another way to find this 100-degree
  angle is to use the exterior-angle rule that follows.

The measure of an exterior angle of a triangle is equal to the sum of the
two nonadjacent interior angles.

Figure 15-7:
Extending
the triangle
to do
calculations.

100°
80°
90°
100 feet
10°
22°

3. Calculate the measure of the third angle.

Adding the two base angles together and subtracting their sum from 180
degrees, you get 180 − (22 + 100) = 180 − 122 = 58 degrees.

4. **Determine the height of the tree.**

- The tree is the side opposite the angle measuring 22 degrees. Using the law of sines, you can write the proportion

$$\frac{\text{tree}}{\sin 22°} = \frac{100}{\sin 58°}$$

- Solve for the height of the tree.

$$\frac{\text{tree}}{\sin 22°} = \frac{100}{\sin 58°}$$

$$\frac{\text{tree}}{0.375} = \frac{100}{0.848}$$

$$(0.375)\frac{\text{tree}}{0.375} = \frac{100}{0.848}(0.375)$$

$$\text{tree} = 44.222$$

The tree is about 44 feet tall.

Continuing with the Law of Cosines

The *law of cosines* comes in handy when you have two or more sides — as in situations involving SSS and SAS — and need the measures of the other three parts. When you have two sides, you need the angle between them. If the angle isn't between the two sides, then you have the ambiguous case, SSA. Although such a situation isn't impossible, you must deal with it carefully. (See the section "Determining a triangle," earlier in this chapter, for more on these cryptic notations.)

Defining the law of cosines

The law of cosines has three different versions that you can use depending on which parts of the triangle you have measures for. Notice the pattern: The squares of the three sides appear in the equations, along with the cosine of the angle opposite one of the sides — the side set equal to the rest of the stuff.

The law of cosines for triangle *ABC* with sides *a*, *b*, and *c* opposite those angles, respectively, says

$$a^2 = b^2 + c^2 - 2bc\cos A$$
$$b^2 = a^2 + c^2 - 2ac\cos B$$
$$c^2 = a^2 + b^2 - 2ab\cos C$$

In plain English, these equations say that the square of one side is equal to the squares of the other two sides, added together, minus twice the product of those two sides times the cosine of the angle opposite the side you're solving for. Whew!

Law of cosines for SAS

When you have two sides of a triangle and the angle between them, you can use the law of cosines to solve for the other three parts. Consider the triangle *ABC* where *a* is 15, *c* is 20, and angle *B* is 124 degrees. Figure 15-8 shows what this triangle looks like.

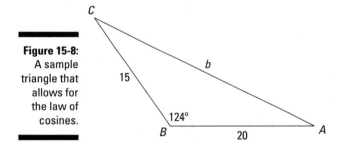

Figure 15-8: A sample triangle that allows for the law of cosines.

Now, to solve for the measure of the missing side and angles:

1. **Find the measure of the missing side by using the law of cosines.**

   Use the law that solves for side *b*.

   $$b^2 = a^2 + c^2 - 2ac \cos B$$
   $$b^2 = 15^2 + 20^2 - 2(15)(20)\cos 124°$$
   $$= 225 + 400 - 600(-0.559)$$
   $$= 960.4$$

   You end up with the value for b^2. Take the square root of each side and just use the positive value (because a negative length doesn't exist).

   $$b^2 = 960.4$$
   $$b = 30.990$$

   The length of side *b* is about 31.

2. **Find the measure of one of the missing angles by using the law of cosines.**

Using the law that solves for a, fill in the values that you know.

$$a^2 = b^2 + c^2 - 2bc \cos A$$
$$15^2 = 31^2 + 20^2 - 2(31)(20) \cos A$$

Solve for cos A by simplifying and moving all the other terms to the left.

$$225 = 961 + 400 - 1240 \cos A$$
$$-1136 = -1240 \cos A$$
$$\frac{-1136}{-1240} = \cos A$$
$$0.916 = \cos A$$

Using Appendix A or a scientific calculator to find angle A, you find that $A = \cos^{-1}(0.916) = 23.652$, or about 24 degrees.

Tip: You can also switch to the law of sines to solve for this angle. Don't be afraid to mix and match when solving these triangles.

3. **Find the measure of the last angle.**

Determine angle B by adding the other two angle measures together and subtracting that sum from 180.

$180 - (124 + 24) = 180 - 148 = 32$. Angle B measures 32 degrees.

How about an application that uses this SAS portion of the law of cosines? Consider the situation: A friend wants to build a stadium in the shape of a regular pentagon (five sides, all the same length) that measures 920 feet on each side. How far is the center of the stadium from the corners? The left part of Figure 15-9 shows a picture of the stadium and the segment you're solving for.

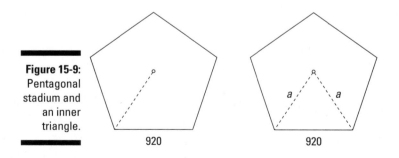

Figure 15-9:
Pentagonal stadium and an inner triangle.

You can divide the pentagon into five isosceles triangles. The base of each triangle is 920 feet, and the two sides are equal, so call them both a. Refer to the right-hand picture in Figure 15-9. Use the law of cosines to solve for a, because you can get the angle between those two congruent sides, plus you already know the length of the side opposite that angle.

1. **Determine the measure of the angle at the center of the pentagon.**

   A circle has a total of 360 degrees. Divide that number by 5, and you find that the angle of each triangle at the center of the pentagon is 72 degrees.

2. **Use the law of cosines with the side measuring 920 feet being the side solved for.**

   $$c^2 = a^2 + a^2 - 2aa\cos C$$
   $$920^2 = 2a^2 - 2a^2\cos 72°$$

   Because the other two sides are the same measure, write them both as a in the equation.

3. **Solve for the value of a.**

   $$920^2 = 2a^2 - 2a^2\cos 72°$$
   $$846,400 = 2a^2(1 - \cos 72°)$$
   $$\frac{846,400}{1 - \cos 72°} = 2a^2$$
   $$\frac{846,400}{1 - 0.309} = 2a^2$$
   $$\frac{846,400}{0.691} = 2a^2$$
   $$1,224,891.462 = 2a^2$$
   $$612,445.731 = a^2$$
   $$782.589 = a$$

   The distance from the center to a corner is between 782 and 783 feet.

Law of cosines for SSS

When you know the values for two or more sides of a triangle, you can use the law of cosines. In the following case, you know all three sides but none of the angles. Solve for the measures of the three angles in triangle *ABC,* which has sides where a is 7, b is 8, and c is 2.

As you can see in Figure 15-10, the triangle appears to have two acute angles and one obtuse angle, the obtuse angle being opposite the longest side.

1. **Solve for the measure of angle A.**

   Using the law of cosines where side a is on the left of the equation, substitute the values that you know and simplify the equation.

Figure 15-10:
A sample
SSS
triangle.

$$a^2 = b^2 + c^2 - 2bc \cos A$$
$$7^2 = 8^2 + 2^2 - 2(8)(2) \cos A$$
$$49 = 64 + 4 - 32 \cos A$$
$$-19 = -32 \cos A$$
$$\frac{-19}{-32} = \cos A$$
$$0.594 = \cos A$$

Now use the table in the Appendix or a scientific calculator to find the measure of A.

$$A = \cos^{-1}(0.594) = 53.559$$

Angle A measures about 54 degrees.

2. **Solve for the measure of angle B.**

Using the law of cosines where side b is on the left of the equation, input the values that you know and simplify the equation.

$$b^2 = a^2 + c^2 - 2ac \cos B$$
$$8^2 = 7^2 + 2^2 - 2(7)(2) \cos B$$
$$64 = 49 + 4 - 28 \cos B$$
$$11 = -28 \cos B$$
$$\frac{11}{-28} = \cos B$$
$$-0.393 = \cos B$$

The negative cosine means that the angle is obtuse — its terminal side is in the second quadrant. Now use the table in the Appendix or a scientific calculator to find the measure of B.

$$B = \cos^{-1}(-0.393) = 113.141$$

Angle B measures about 113 degrees.

3. **Determine the measure of angle *C*.**

   Because angle A measures 54 degrees and angle B measures 113 degrees, add them together and subtract the sum from 180 to get the measure of angle C.

   $180 - (54 + 113) = 180 - 169 = 11$. Angle C measures only 11 degrees.

Being ambiguous

Many people are visual learners, solving problems better by using a picture. This characteristic will serve such people well when it comes to solving triangles that are SSA, meaning that they know the measures of two sides and an angle that isn't between those sides. Drawing a picture helps explain why the situation may have more than one answer. When you use this setup in an actual application, the correct answer is usually pretty clear. First, I show you how to do one of these problems in general and then I show how it may actually play out in real life.

Find the missing parts of the triangle ABC that has sides a and b measuring 85 and 93, respectively, and angle A measuring 61 degrees. Figure 15-11 presents the situation.

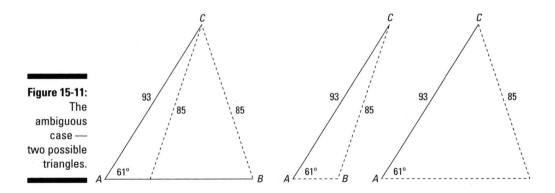

Figure 15-11:
The ambiguous case — two possible triangles.

1. **Find the length of side *c* by using the law of cosines with *a* on the left-hand side of the equation.**

   Use this form because after you input the known values, it's the only one that will have just one variable to solve for — even though that variable has two powers.

- Enter the values into the law of cosines.

$$a^2 = b^2 + c^2 - 2bc \cos A$$
$$85^2 = 93^2 + c^2 - 2(93)(c) \cos 61°$$

- Simplify the equation by performing all the operations and getting the variables alone on the right side.

$$7225 = 8649 + c^2 - 186(c)0.485$$
$$-1424 = c^2 - 90.21c$$

You end up with a quadratic equation.

- Use either the quadratic formula or a calculator to determine the solutions.

$$0 = c^2 - 90.21c + 1424$$
$$c = 69.813 \text{ or } 20.397$$

So c measures either 70 or 20.

2. **Let c measure 70, and find the measures of the other two angles.**

This time, take a departure from the law of cosines and use, instead, the law of sines.

- Use angle A and side a, and pair the ratio with angle C and side c to get

$$\frac{\sin A}{a} = \frac{\sin C}{c}$$
$$\frac{\sin 61°}{85} = \frac{\sin C}{70}$$

- Now multiply each side by 70, and solve for the sine of C.

$$70 \cdot \frac{\sin 61°}{85} = \frac{\sin C}{\cancel{70}} \cdot \cancel{70}$$
$$70 \cdot \frac{0.875}{85} = \sin C$$
$$0.721 = \sin C$$

- Solve for the angle with that sine.

$$C = \sin^{-1}(0.721) = 46.137$$

The measure of angle C is about 46 degrees.

If angle A is 61 degrees and angle C is 46 degrees, then angle B is 180 degrees minus the sum of A and C: $180 - (61 + 46) = 180 - 107 = 73$ degrees.

3. **Now let c measure 20, and find the measures of the other two angles.**

Go back to the law of cosines to do this part. You can compare the two methods — the one in this step and the one in Step 2 — to see which one you like better.

- Use the law with c on the left-hand side of the equation to solve for the cosine of angle C.

$$c^2 = a^2 + b^2 - 2ab\cos C$$
$$20^2 = 85^2 + 93^2 - 2(85)(93)\cos C$$
$$400 = 7225 + 8649 - 15,810\cos C$$
$$\frac{-15,474}{-15,810} = \cos C$$
$$0.979 = \cos C$$

- Use the table in the Appendix or a calculator to find the measure of angle C.

$$C = \cos^{-1}(0.979) = 11.763°$$

Angle C measures about 12 degrees, which means that angle B is 180 – (61 + 12) = 180 – 73 = 107 degrees.

The ambiguous case causes a bit of confusion. Why would you want two answers? The following example may help clear up this mystery. You really *don't* want two answers. You just want the one that answers your question.

Slim and Jim are both sitting at the intersection of two roads, which forms a 50-degree angle. They leave the intersection at the same time — Slim in his old, slow, beat-up pickup truck, and Jim in his nifty-swifty Jeep. When Jim was 400 yards down the road, the two of them were 320 yards apart. How far had Slim driven at that point?

You definitely need a picture for this problem. See Figure 15-12.

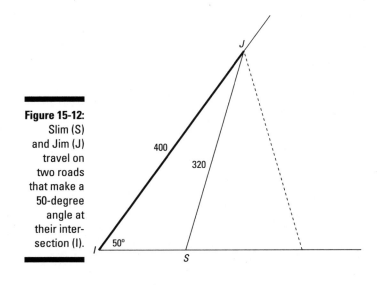

Figure 15-12: Slim (S) and Jim (J) travel on two roads that make a 50-degree angle at their intersection (I).

Two wrongs make a right

When students are first introduced to fractions, they're often tempted to take some liberties with the rules that can get them into trouble. Imagine the frustration to the teacher and student alike when the student stumbles on one of the four fractions, with two digits in the numerator and denominator, where incorrect cancellation results in a correct answer. The four fractions where such a situation can occur are $\frac{64}{16}$, $\frac{98}{49}$, $\frac{95}{19}$, and $\frac{65}{26}$.

When a student mistakenly crosses out the two like digits, the result is actually the correct answer: $\frac{\cancel{6}4}{1\cancel{6}} = \frac{4}{1} = 4$, $\frac{\cancel{9}8}{4\cancel{9}} = \frac{8}{4} = 2$, $\frac{\cancel{9}5}{1\cancel{9}} = \frac{5}{1} = 5$, and $\frac{\cancel{6}5}{2\cancel{6}} = \frac{5}{2}$. Thank goodness only four such fractions exist.

You can safely assume that Slim couldn't have gone farther than Jim in his old clunker — unless his truck had hidden powers. Figure out how far Slim drove, the distance from I to S (in this example, I call the distance j to be consistent with the triangle labels) by using the law of cosines. The side s is 400 yards, and angle I is 50 degrees.

1. **Write the law of cosines, and replace the letters with the values.**

$$i^2 = s^2 + j^2 - 2(s)(j)\cos I$$
$$320^2 = 400^2 + j^2 - 2(400)(j)\cos 50°$$
$$102,400 = 160,000 + j^2 - 800j(0.643)$$
$$0 = 57,600 + j^2 - 514.4j$$

This equation simplifies to a quadratic equation with the variable j.

2. **Solve the quadratic equation.**

$$0 = j^2 - 514.4j + 57,600$$

Use a calculator or the quadratic formula, and you get two solutions: $x = 349.676$ and $x = 164.723$. Either answer gives you a distance smaller than the distance that Jim traveled. Refer to Figure 15-12, and choose the answer that appears to be correct.

Including the Law of Tangents

The law of tangents isn't nearly as popular as the law of sines or the law of cosines, but leaving it out just doesn't seem fair. You use this law with SAS — when you have two sides and the angle between them. The law of tangents may look a bit intimidating, but it's really quite nice.

In triangle ABC, with sides a, b, and c opposite the respective angles A, B, and C, the law of tangents states

$$\frac{a-b}{a+b} = \frac{\tan\frac{1}{2}(A-B)}{\tan\frac{1}{2}(A+B)}$$

$$\frac{b-c}{b+c} = \frac{\tan\frac{1}{2}(B-C)}{\tan\frac{1}{2}(B+C)}$$

$$\frac{c-a}{c+a} = \frac{\tan\frac{1}{2}(C-A)}{\tan\frac{1}{2}(C+A)}$$

The best way to show how the law of tangents works is with an example. In triangle ABC, a is 52, b is 28, and angle C is 80 degrees. To find the remaining parts of the triangle:

1. **Use the law of tangents involving sides a and b.**

   Fill in the values that you know and simplify.

   $$\frac{a-b}{a+b} = \frac{\tan\frac{1}{2}(A-B)}{\tan\frac{1}{2}(A+B)}$$

   $$\frac{52-28}{52+28} = \frac{\tan\frac{1}{2}(A-B)}{\tan\frac{1}{2}(A+B)}$$

   $$\frac{24}{80} = \frac{3}{10} = \frac{\tan\frac{1}{2}(A-B)}{\tan\frac{1}{2}(A+B)}$$

2. **Multiply each side by the denominator on the right.**

   $$\tan\frac{1}{2}(A+B) \cdot \frac{3}{10} = \frac{\tan\frac{1}{2}(A-B)}{\tan\frac{1}{2}(A+B)} \cdot \tan\frac{1}{2}(A+B)$$

   $$\tan\frac{1}{2}(A+B) \cdot \frac{3}{10} = \tan\frac{1}{2}(A-B)$$

3. **Determine $A + B$.**

   The sum of angles A and B equals 180 degrees minus the measure of angle C: $A + B = 180 - 80 = 100$.

4. **Replace $(A + B)$ with 100 and simplify.**

   Use the table in the Appendix or a scientific calculator to do the calculations.

$$\tan \tfrac{1}{2}(100) \cdot \frac{3}{10} = \tan \tfrac{1}{2}(A - B)$$

$$\tan 50° \cdot \frac{3}{10} = \tan \tfrac{1}{2}(A - B)$$

$$1.192 \cdot \frac{3}{10} = \tan \tfrac{1}{2}(A - B)$$

$$0.3576 = \tan \tfrac{1}{2}(A - B)$$

5. **Find the value of $A - B$.**

$$\tfrac{1}{2}(A - B) = \tan^{-1}(0.3576) = 19.677$$

$$\tfrac{1}{2}(A - B) = 19.677$$

$$A - B = 39.354$$

The difference between angle A and angle B is 39.354, or about 39 degrees.

6. **Determine A and B by solving the system of equations for $A + B$ and $A - B$.**

Eliminate one of the variables by adding the two equations together.

$$A + B = 100$$
$$\underline{A - B = 39}$$
$$2A \quad\; = 139$$
$$A \quad = \frac{139}{2} = 69.5$$

Because A is 69.5, B equals 100 minus 69.5, or 30.5. Rounding these values to whole numbers, you get $A = 70$ and $B = 30$.

7. **Solve for side c by using the law of sines.**

$$\frac{c}{\sin C} = \frac{a}{\sin A}$$

$$\frac{c}{\sin 80°} = \frac{52}{\sin 70°}$$

$$\frac{c}{0.985} = \frac{52}{0.940}$$

$$c = \frac{52}{0.940}(0.985) = 54.489$$

Side c is about 54 units long.

Finding the Areas of Triangles

Finding the area of a triangle sounds relatively easy. Most grade-school children get plenty of chances to do just that. They're given a triangle and the length of the base and the height, or altitude, drawn to that base. Simple! Just plug those values into the formula, and you have it. But think about it: How many times do you have a triangular plot of land or triangular sail for a boat and have the measure of the *altitude?*

What you find in this section is a formula for every occasion. Give me a triangle, and I can find the area. Although the base and altitude would be nice, I can also do the problem with the measures of the three sides. You have two sides and an included angle? Sure, I can do that. How about two angles and an included side? I have a formula for that, too.

Of course, if you don't have the measurements for one of these exact situations, you can go to the law of sines, cosines, or tangents to fill in the blanks and find the sides or angles that you need.

Finding area with base and height

The most basic formula for finding the area of a triangle occurs when you know the base and the height. The height is drawn perpendicular to the base up to the vertex opposite that base. Figure 15-13 shows you what I mean.

The equation for the area *(A)* of a triangle with base *b* and height *h* is $A = \frac{1}{2}bh$.

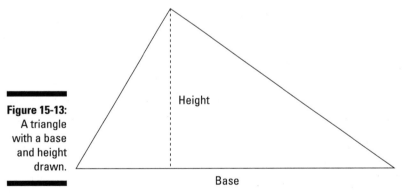

Figure 15-13:
A triangle with a base and height drawn.

Height

Base

For example, to find the area of a triangle with a base measuring 12 inches and a height measuring 5 inches, input the values into the equation. You find that $A = \frac{1}{2}(12)(5) = 30$, or 30 square inches.

If the triangle happens to be a right triangle, then you're really in business. The base and height are the legs, or the two sides that are perpendicular to one another. Just find half of the base times the height. Here's an example.

Kirsten has a corner lot and wants to make a triangular garden where the two sidewalks meet. She has a 20-foot piece of border to go along the diagonal, or hypotenuse, of the triangle. She wants one side along the sidewalk to be 12 feet. How many square feet of garden will she have? Figure 15-14 illustrates the situation.

Figure 15-14:
Kirsten's
triangular
garden.

1. **Find the length of the other leg of the right triangle.**

   Using the Pythagorean theorem, and calling the missing length x, you get

   $x^2 + 12^2 = 20^2$

   $x^2 + 144 = 400$

   $x^2 = 400 - 144 = 256$

   $x = \sqrt{256} = 16$

   The other side is 16 feet long.

2. **Find the area of the triangle.**

   The base is 12 feet, and the height is 16 feet. Using the formula, you get

   $A = \frac{1}{2}bh = \frac{1}{2}(12)(16) = 96$

   The area is 96 square feet, which is a lot of garden to weed!

Finding area with three sides

Suppose that you have 240 yards of fencing, and you decide to build a triangular corral for your llama. Why triangular? You heard that llamas favor the shape, of course. You want the llama to have enough room to run around, so you need to know the area. What should the lengths of the triangle's sides be? You can solve this little problem by using Heron's Formula for the area of a triangle.

Heron's formula says that if a triangle *ABC* has sides of lengths *a, b,* and *c* opposite the respective angles, and you let the semiperimeter, *s,* be half of the triangle's perimeter, then the area of the triangle is $A = \sqrt{s(s-a)(s-b)(s-c)}$.

In the problem of the fencing and the llama, you have many ways to make a triangular corral from 240 yards of fencing. Figure 15-15 shows a few of the possibilities. Notice that in each case, the lengths of the sides add up to 240. For the sake of this problem, don't worry about a gate.

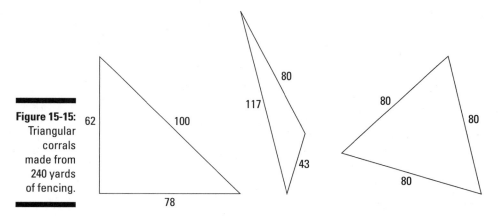

Figure 15-15: Triangular corrals made from 240 yards of fencing.

Which triangle has the greatest area? Obviously, one of them is a bit on the scrawny side, even though it uses up 240 yards of fencing, like the others. Here's how to compute the areas for the three triangles.

1. **Find the semiperimeter, *s*, for each triangle.**

   Referring to Figure 15-15:

   - Left triangle: $\frac{1}{2}(62 + 100 + 78) = \frac{1}{2}(240) = 120$

   - Center triangle: $\frac{1}{2}(117 + 80 + 43) = \frac{1}{2}(240) = 120$

   - Right triangle: $\frac{1}{2}(80 + 80 + 80) = \frac{1}{2}(240) = 120$

   Not surprisingly, all the semiperimeters are the same, because all the perimeters are 240.

2. Use Heron's Formula to find each area.

Again, referring to Figure 15-15:

- Left triangle: $A = \sqrt{120(120 - 62)(120 - 100)(120 - 78)} = \sqrt{120(58)(20)(42)} = 2417.933$
- Center triangle: $A = \sqrt{120(120 - 117)(120 - 80)(120 - 43)} = \sqrt{120(3)(40)(77)} = 1052.996$
- Right triangle: $A = \sqrt{120(120 - 80)(120 - 80)(120 - 80)} = \sqrt{120(40)(40)(40)} = 2771.281$

The triangle on the right has the greatest area. Of the shapes in Figure 15-15, that triangle is the best. But you may be wondering whether another shape gives more area than that one. The answer: no. With calculus, you can prove that an equilateral triangle gives you the greatest possible area with any amount of fencing. Without calculus, you just have to try a bunch of shapes to convince yourself.

Finding area with SAS

When you know the lengths of two of a triangle's sides plus the measure of the angle between those sides, you can find the area of the triangle. This method requires a little trigonometry — you have to find the sine of the angle involved. But the formula is really straightforward.

If triangle ABC has sides measuring a, b, and c opposite the respective angles, then you can find the area with one of these formulas:

$$A = \frac{1}{2}ab\sin C, \ A = \frac{1}{2}bc\sin A, \text{ or } A = \frac{1}{2}ac\sin B$$

For example, look at the 30-60-90 right triangle in Figure 15-16. I use this particular example because the numbers come out so nicely.

First, find the area by using angle B and the two sides forming it.

1. Choose the correct version of the formula.

The formula that uses angle B is $A = \frac{1}{2}ac\sin B$.

2. Find the sine of the angle.

$$\sin 60° = \frac{\sqrt{3}}{2}$$

3. Substitute the values into the formula and simplify.

$$A = \frac{1}{2}\left(\cancel{18}^{9}\right)(9)\frac{\sqrt{3}}{2} = \frac{81\sqrt{3}}{2}$$

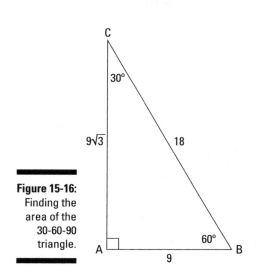

Now find the area by using angle C and the two sides forming it.

1. **Choose the correct version of the formula.**

   The formula that uses angle C is $A = \frac{1}{2}ab\sin C$.

2. **Find the sine of the angle.**

   $\sin 30° = \frac{1}{2}$

3. **Substitute the values into the formula and simplify.**

   $$A = \frac{1}{2}\left(\cancel{18}^{9}\right)\left(9\sqrt{3}\right)\frac{1}{2} = \frac{81\sqrt{3}}{2}$$

Using the method involving angle A gives you the same result, of course. For a quick comparison, just use the formula $A = \frac{1}{2}bh$ to find the area, because you're dealing with a right triangle: $A = \frac{1}{2}(9)\left(9\sqrt{3}\right) = \frac{81\sqrt{3}}{2}$. The methods all produce the same result.

Finding area with ASA

As you probably suspected, when you have two angles and the side between them, you can find the area of a triangle. The formulas go as follows.

In triangle ABC, if the measures of the sides are a, b, and c opposite the respective angles, then you can determine the area by using one of the following equations:

$$A = \frac{a^2 \sin B \sin C}{2 \sin A}, \quad A = \frac{b^2 \sin A \sin C}{2 \sin B}, \quad \text{or} \quad A = \frac{c^2 \sin A \sin B}{2 \sin C}$$

These formulas are actually built from the formula for finding the area with SAS, with a little help from the law of sines. Here's how one of them came to be.

1. **Start with the SAS rule for area.**

   $$A = \frac{1}{2} ab \sin C$$

2. **Write the law of sines involving angles *A* and *B*.**

   $$\frac{a}{\sin A} = \frac{b}{\sin B}$$

3. **Solve for *b*.**

   $$\sin B \cdot \frac{a}{\sin A} = \frac{b}{\cancel{\sin B}} \cdot \cancel{\sin B}$$

   $$\frac{\sin B \cdot a}{\sin A} = b$$

4. **Substitute the equivalent for *b* into the area formula.**

   $$A = \frac{1}{2} a \left(\frac{\sin B \cdot a}{\sin A} \right) \sin C$$

   $$A = \frac{a^2 \sin B \sin C}{2 \sin A}$$

Now consider an example. Say you have a triangle with angles *A*, which is 45 degrees, and *B*, which is 55 degrees, and the side between them, *c*, equal to 10. Find the area.

1. **Choose the correct formula — the one with c^2 in it.**

   $$A = \frac{c^2 \sin A \sin B}{2 \sin C}$$

2. **Find the sines of the two given angles.**

   The sine of 45 degrees equals 0.707, and the sine of 55 degrees equals 0.819.

3. **Find the sine of the third angle.**

   Angle *C* measures 180 − (45 + 55), or 180 − 100, which equals 80 degrees.

   The sine of 80 degrees equals 0.985.

4. **Substitute the values into the formula and solve.**

   $$A = \frac{10^2 (0.707)(0.819)}{2(0.985)} = 29.393$$

The area is a little over 29 square units.

Part V
The Graphs of Trig Functions

The 5th Wave   By Rich Tennant

I'm mathematically dyslexic. But it's not that unusual - 100 out of every 15 people are.

In this part . . .

A picture is worth a thousand words, or so they say. The graphs of the trigonometric functions certainly qualify here. In that case, this part is theoretically about 1,000 pages long, because I provide all the graphs you need to see the similarities and differences among the six basic trig functions, their variations, and the inverse trig functions.

Chapter 16

Graphing Sine and Cosine

- -

In This Chapter

▶ Looking at the basic graphs of sine and cosine

▶ Working with variations of the graphs

▶ Using sine and cosine curves to make predictions

- -

*T*he graphs of the sine and cosine functions are very similar. If you look at them without a coordinate axis for reference, you can't tell them apart. They keep repeating the same values over and over — and the values, or outputs, are the same for the two functions. These two graphs are the most recognizable and useful for modeling real-life situations. The sine and cosine curves can represent anything tied to seasons — the weather, shopping, hunting, and daylight. The equations and graphs of the curves are helpful in describing what happens during those seasons.

The ABCs of Graphing

You can graph trig functions in a snap — well, maybe not that fast — but you can do it quickly and efficiently with just a few pointers. If you set up the axes properly and have a general knowledge of the different functions' shapes, then you're in business.

Different kinds of values represent the two axes in trig graphs. The *x*-axis is in angle measures, and the *y*-axis is in plain old numbers. The *x*-axis is labeled in either degrees or radians. Often, a graph represents the values from -2π to 2π to accommodate two complete cycles of the sine, cosine, secant, or cosecant functions (or four complete cycles of the tangent or cotangent functions). If the *x*-axis is labeled in degrees, it typically ranges from -360 to 360, which is a wide number range. That range is in sharp contrast to the *y*-axis, which often just goes from -5 to 5. You'll find that radians — which are real numbers — are preferable. The *y*-axis is labeled in real numbers; how high and low the range extends depends on the particular function or variation of a function that you're graphing.

If you're using a graphing calculator, you need to be aware of what mode you're in when creating graphs. Otherwise, you'll get completely baffling results or none at all. For more on changing your calculator's mode, head on back to Chapter 13. It just takes the press of a button or two, and you're in the right mood — oops, mode.

Waving at the Sine

The graph of the sine function is a nice, continuous wave that rolls along gently and keeps repeating itself. The domain, or *x*-values, of the sine function includes all angles in degrees or all real numbers in radians, so the curve has no breaks or holes. The range, or *y*-values, of the sine function consists of all the numbers between –1 and 1, including those two values. Figure 16-1 shows a graph of the sine function from about –2π to 2π (or from about –360 to 360 degrees).

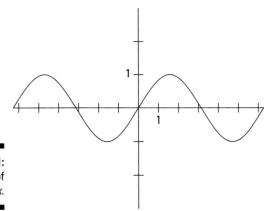

Figure 16-1:
The graph of
y = sin *x*.

Figure 16-1 shows two complete cycles of the sine curve — the curve goes through its routine twice on the graph. If you could see the sine curve forever in either direction, it wouldn't look any different. The curve repeats the same pattern over and over again, to infinity and beyond.

Describing amplitude and period

The sine function and any of its variations have two important characteristics: the amplitude and period of the curve. You can determine these characteristics by looking at either the graph of the function or its equation.

Gaining height with the amplitude

The *amplitude* of the sine function is the distance from the middle value or line running through the graph up to the highest point. In other words, the amplitude is half the distance from the lowest value to the highest value. In the sine and cosine equations, the amplitude is the coefficient (multiplier) of the sine or cosine. For example, the amplitude of $y = \sin x$ is 1. To change the amplitude, multiply the sine function by a number. Take a look at Figure 16-2, which shows the graphs of $y = 3\sin x$ and $y = \frac{1}{2} \sin x$.

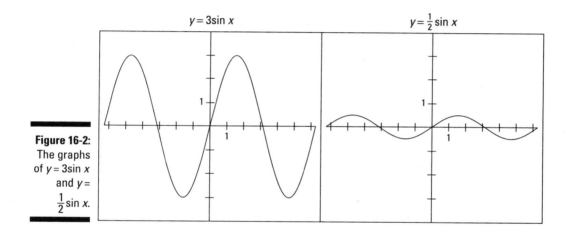

Figure 16-2: The graphs of $y = 3\sin x$ and $y = \frac{1}{2}\sin x$.

As you can see, multiplying by a number greater than 1 makes the graph extend higher and lower. The amplitude of $y = 3\sin x$ is 3. Conversely, multiplying by a number smaller than 1 (but bigger than 0) makes the graph shrink in value — it doesn't go up or down as far. The amplitude of $y = \frac{1}{2}\sin x$ is $\frac{1}{2}$.

The sound of music

Sounds are created by vibrations. Tuning forks can produce pure tones when they vibrate, and sine waves can model those tones. A formula for a pure tone is $y = A \sin(2\pi f t)$, where A stands for amplitude (loudness), f is for frequency (vibrations per second), and t is a unit of time. If a string, tuning fork, or something similar vibrates at the rate of 256 times per second, then you hear middle C. When you double the frequency of any pure tone, you go up one octave, so 512 vibrations per second gives you the C above middle C.

If you add waves of different frequencies and loudness together, you get more-interesting and complex tones. The string of a violin or the inside of an oboe, for example, can vibrate with more than one frequency at the same time.

Punctuating with the period

The *period* of a function is the extent of input values it takes for the function to run through all the possible values and start all over again in the same place to repeat the process. In the case of the sine function, the period is 2π, or 360 degrees. Pick any place on the sine curve, follow the curve to the right or left, and 2π or 360 units from your starting point along the *x*-axis, the curve starts the same pattern over again.

Multiplying the angle variable, *x*, by a number changes the period of the sine function. If you multiply the angle variable by 3, such as in $y = \sin 3x$, then the curve will make three times as many completions in the usual amount of space. So multiplying by 3 actually *reduces* the length of the period. In the case of $y = \sin \frac{1}{2}x$, only half of the curve fits in the same space. So a coefficient less than 1 increases the number of inputs that the function needs to complete a cycle. Figure 16-3 shows pictures of these two graphs.

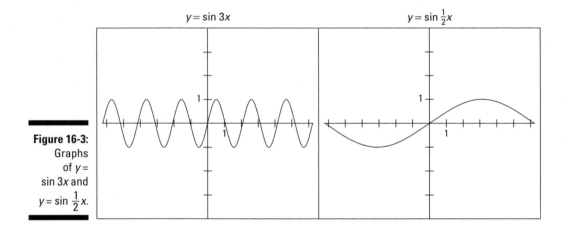

$$y = \sin 3x \qquad\qquad y = \sin \tfrac{1}{2}x$$

Figure 16-3: Graphs of $y = \sin 3x$ and $y = \sin \frac{1}{2}x.$

The location of the multiplier makes a big difference. Multiplying the sine function by 4 and its angle variable by 4 results in two completely different graphs. The graph of $y = 4\sin x$ is much higher than usual — the amplitude is greater than that of the standard sine function. The graph of $y = \sin 4x$ has an amplitude of 1, but the period is smaller and the curve is more scrunched together — it repeats over and over more quickly.

Formalizing the sine equation

A general equation for the sine function is $y = A \sin Bx$. The *A* and *B* are numbers that affect the amplitude and period of the basic sine function, respectively.

TRIG RULES
$\frac{1}{+1}$
$\frac{}{2}$

The graph of the function $y = A \sin Bx$ has an amplitude of A and a period of $\frac{2\pi}{B}$. The amplitude, A, is the distance measured from the y-value of a horizontal line drawn through the middle of the graph (or the average value) to the y-value of the highest point of the sine curve, and B is the number of times the sine curve repeats itself within 2π, or 360 degrees.

By keeping these two values in mind, you can quickly sketch the graph of a sine curve — or picture it in your head. For example, when graphing $y = 4\sin 2x$:

1. **Adjust for the amplitude.**

   The amplitude is 4, so the curve will extend up 4 units and down 4 units from the middle. To allow for some space above and below, set the y-axis to go from –5 to 5.

2. **Take into account the period.**

   Two complete graphs of the sine are within the space that usually houses only one.

3. **Graph the curve from -2π to 2π (see Figure 16-4).**

   You can see that the graph goes from –4 to 4 and that four complete cycles are in the space that usually houses only two.

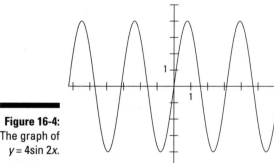

Figure 16-4:
The graph of
$y = 4\sin 2x$.

Translating the sine

Playing around with the amplitude and period of the sine curve can result in some interesting changes to the basic curve. That curve is still recognizable, though. You can see the rolling, smooth curve crossing back and forth over a middle line. In addition to those changes, you have two other options for altering the sine curve — shifting the curve up, down, or sideways. These shifts are called *translations* of the curve. (Flip back to Chapter 3 for a basic discussion on translating functions.)

Sliding up or down

You can move a sine curve up or down by simply adding or subtracting a number from the equation of the curve. For example, the graph of $y = \sin x + 4$ moves the whole curve up 4 units, with the sine curve crossing back and forth over the line $y = 4$. On the other hand, the graph of $y = \sin x - 1$ slides everything down 1 unit. Figure 16-5 shows what the two graphs look like.

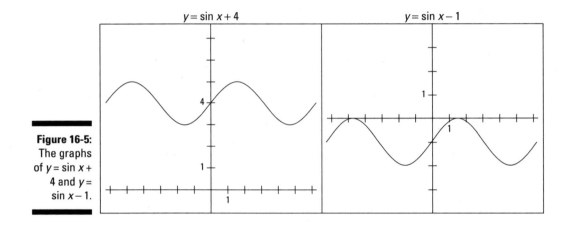

Figure 16-5: The graphs of $y = \sin x + 4$ and $y = \sin x - 1$.

As you can see, the basic shape of the sine curve is still recognizable — the curves are just shifted up or down on the coordinate plane.

Shifting left or right

By adding or subtracting a number from the angle in a sine equation, you can move the curve to the left or right of its usual position. This shift, or translation, relates the sine curve to the cosine curve. But the translation of the sine itself is important: Shifting the curve left or right can change the places that the curve crosses the *x*-axis or some other horizontal line. For example, the graph of $y = \sin (x + 1)$ is the usual sine curve slid 1 unit to the left, and the graph of $y = \sin (x - 3)$ slides 3 units to the right. Figure 16-6 shows the graphs of the original sine equation and these two shifted equations.

Take a look at the point marked on each graph in Figure 16-6. This point illustrates how an *intercept* (where the curve crosses an axis) shifts on the graph when you add or subtract a number from the angle variable.

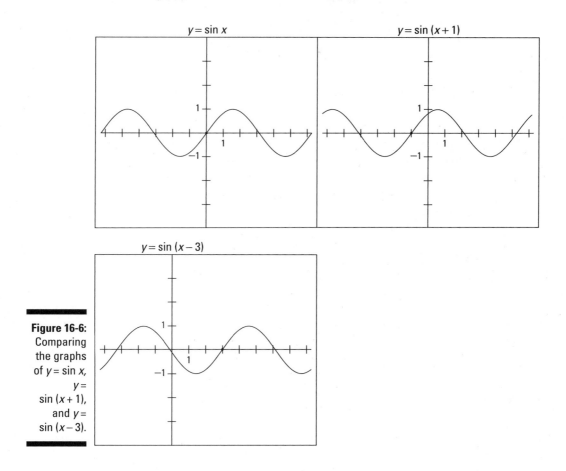

$y = \sin x$

$y = \sin (x + 1)$

$y = \sin (x - 3)$

Figure 16-6:
Comparing
the graphs
of $y = \sin x$,
$y =$
$\sin (x + 1)$,
and $y =$
$\sin (x - 3)$.

Note the difference between adding or subtracting a number to the function and adding or subtracting a number to the angle measure. These operations affect the curve differently, as you can see by comparing Figure 16-5 and Figure 16-6.

$y = \sin x + 2$ Adding 2 to the function raises the curve by 2 units.

$y = \sin (x + 2)$ Adding 2 to the angle variable shifts the curve 2 units to the left.

Graphing Cosine

The graph of the cosine function looks very much like that of the sine function. This quality is due to the fact that they're related by domain and range and also by several identities. An identity involving a shift explains the relationship best, because that shift can make the graph of the sine function look like the cosine function.

Comparing cosine to sine

The relationship between the sine and cosine graphs is that the cosine is the same as the sine shifted to the left by 90 degrees, or $\frac{\pi}{2}$. The equation that represents this relationship is $\cos x = \sin\left(x + \frac{\pi}{2}\right)$. Look at the graphs of the sine and cosine functions on the same coordinate axes, as shown in Figure 16-7. The graph of the cosine is the darker curve; note how it's shifted to the left of the sine curve.

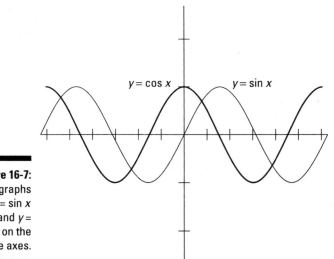

Figure 16-7: The graphs of $y = \sin x$ and $y = \cos x$ on the same axes.

The graphs of the sine and cosine functions illustrate a property that exists for several pairings of the functions. This property is based on the right triangle and the two acute or complementary angles in a right triangle. The identities that arise from the triangle are called the *cofunction identities*.

The cofunction identities are

$$\sin\theta = \cos(90° - \theta) \qquad \csc\theta = \sec(90° - \theta)$$
$$\cos\theta = \sin(90° - \theta) \qquad \sec\theta = \csc(90° - \theta)$$
$$\tan\theta = \cot(90° - \theta) \qquad \cot\theta = \tan(90° - \theta)$$

These identities show how the function values of the complementary angles in a right triangle are related. For example, $\cos\theta = \sin(90° - \theta)$ means that if θ is equal to 25 degrees, then $\cos 25° = \sin(90° - 25°) = \sin 65°$. This equation is a roundabout way of explaining why the graphs of sine and cosine are different by just a slide. You probably noticed that these cofunction identities all use the difference of angles, but the slide of the sine function to the left was a sum. The shifted sine graph and the cosine graph are really equivalent — they become graphs of the same set of points. Here's how to prove it.

1. **Write an equation with the sine slid to the left set equal to the cofunction identity involving cosine.**

$$\sin\left(x + \frac{\pi}{2}\right) = \sin\left(\frac{\pi}{2} - x\right)$$

2. **Apply the two identities for the sine of the sum and difference of two angles (see Chapter 11).**

The two identities are

$$\sin(a + b) = \sin a \cos b + \cos a \sin b$$
$$\sin(a - b) = \sin a \cos b - \cos a \sin b$$

Substituting in the x's and angles,

$$\sin x \cos\frac{\pi}{2} + \cos x \sin\frac{\pi}{2} = \sin\frac{\pi}{2}\cos x - \cos\frac{\pi}{2}\sin x$$

3. **Simplify the terms by using the values of the functions.**

$$\sin x(0) + \cos x(1) = (1)\cos x - (0)\sin x$$
$$\cos x = \cos x$$

So you see, the shifted sine graph is equal to the cosine graph.

Using properties to graph cosine

The cosine function has the same amplitude and period as the sine function: The amplitude is 1, and the period is 2π, or 360 degrees. The variations on the cosine work the same way as on the sine. If you want to change the amplitude, multiply the cosine function by a number. If you want to change the period, multiply or divide the angle variable by a number. To slide the whole curve up, down, right, or left, add or subtract a number from the whole function or the angle variable. Here are some examples.

For the equation $y = 3\cos x$, the amplitude is 3, meaning the graph stretches up and down to 3 units. The graph of $y = \cos(x - 3)$ is shifted 3 units to the right. Figure 16-8 shows you the graphs.

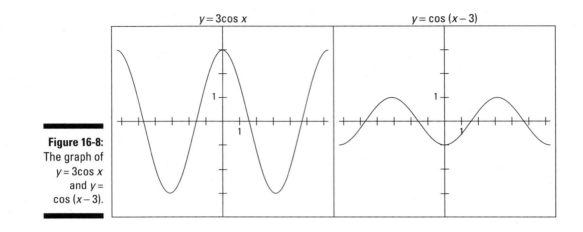

Figure 16-8:
The graph of
$y = 3\cos x$
and $y = \cos(x - 3)$.

The graphs of sine and cosine are difficult to tell apart when they're shifted about. But that fact just shows how much those functions have in common, which can work to your advantage when you're applying them.

Applying the Sines of the Times

The sine curve and its cofunction, cosine, are great for modeling situations that happen over and over again in a predictable fashion. Some examples include the weather, seasonal sales of goods, body temperature, the tide's height in a harbor, average temperatures, and so on. In this section, I show you a few examples of how you can use these functions in practical situations. In each case, I point out how the graph and formula illustrate the amplitude, period, and any shifts (for more on those concepts, check out the "Waving at the Sine" section, earlier in this chapter).

Sunning yourself

San Diego, California, is a gorgeous part of the world. Whether it's summer or winter, you want to be there. But what if you're someone who likes long, sunny days? When is the best time to go there? Assume that the following formula gives you the number of hours of daylight in San Diego when you input any day of the year. Letting t be the day of the year (from 1 to 365), you can figure the number of hours of sunlight, H, with the equation $H(t) = 2.4\sin(0.017t - 1.377) + 12$. Figure 16-9 shows the graph of this equation.

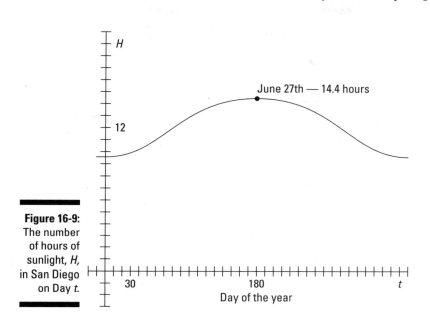

Figure 16-9:
The number
of hours of
sunlight, *H,*
in San Diego
on Day *t.*

The amplitude of the sine curve is 2.4, which means that the number of daylight hours extends 2.4 hours above and below the average number of daylight hours. The average number of daylight hours is 12, which is the translation upward, meaning that the hours of sunlight range from 14.4 to 9.6, depending on the time of year. The period is $\frac{2\pi}{0.017} \approx 370$, which is a little longer than a year because of the rounding in the formula. The coefficient on the *t* means that 0.017 of the curve takes up the usual amount of space for one curve, 2π units. As you can see on the graph, the day with the most sunlight is June 27. You can determine that high point by using a graphing calculator that finds it for you — as well as the *y*-value of when *t* equals 14.4, or you can use calculus to solve it! Do you know, now, when you want to go to San Diego?

Trusty old protractors

A protractor is a familiar instrument to grade-school and high-school students. They use this flat, semicircular instrument, which is marked with degrees from 0 to 180 degrees, to construct and measure angles.

The protractor has been around for a very long time. The first protractors were used in navigation to plot the positions of ships on navigational charts. In 1801, Joseph Huddart, a U.S. Navy captain, invented an instrument called a *three-arm protractor,* or *station pointer.*

Averaging temperature

A relatively reasonable model for the average daily temperature in Peoria, Illinois, is $T(x) = 50 - 42\cos(0.017x - 0.534)$, where x is the day of the year starting with January 1 as Day 1. The $T(x)$ represents the temperatures in degrees Fahrenheit. The graph is in radian measure, because radians are real numbers, as opposed to degrees — you need numbers to count off the days. Figure 16-10 shows what the graph of the function looks like for the whole year.

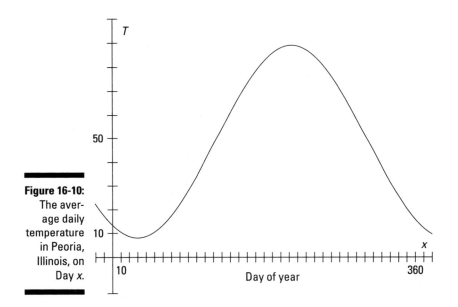

Figure 16-10: The average daily temperature in Peoria, Illinois, on Day x.

The multiplier on the cosine function is 42, so the amplitude of the curve is 42. Don't worry about the negative sign in front of the 42. The curve goes upward and downward anyway, so the negative sign just makes it go downward and then upward, instead of the reverse.

The period is affected by the multiplier 0.017. The result of that multiplication is that only 0.017 of a cosine curve takes up the usual amount of space for an entire curve, which is 2π, or a little over 6 units. Because this graph is for a whole year, the curve has to spread out over 365 units, so that each of the horizontal units has just a little part of it.

The shift upward of 50 units is the middle temperature for the year. Add the amplitude of 42 to this number, and the average temperature gets up to 92 degrees; subtract the amplitude, and the average gets down to 8 degrees. Note that the curve starts a little to the right of the y-axis to account for when the seasons change. If you want more details on curve translations to the left, right, up, and down, go to Chapter 19.

What do you do with the graph? You can estimate when the highest and lowest temperatures occur and get an idea of the types of temperatures to expect if you move to Peoria, Illinois. Figure 16-11 shows the graph of the average temperatures with points for some days of the year and the average temperatures on those days. A graphing calculator is indispensable when graphing these figures and calculating values. You can either input *x*-values to find out what the *y*-values are at those points, or you can trace along the curve to get the measures.

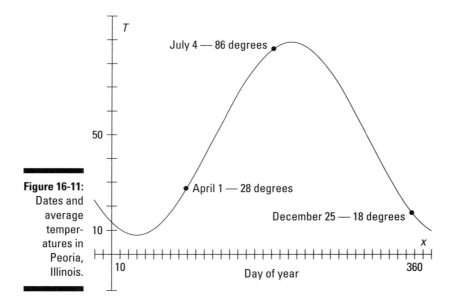

Figure 16-11:
Dates and average temperatures in Peoria, Illinois.

Taking your temperature

The temperature of a person's body fluctuates during the day instead of staying at a normal 98.6 degrees. And actually, not everyone has a "normal" temperature. Lots of people run either hot or cold.

If you're one of the special people with a normal temperature, then your temperature goes up and down by about 1 degree each day. The formula $T(x) = \sin(x + 0.262) + 98.6$ may be a model of your temperature during a 24-hour period. The variable *x* is the number of hours since midnight, so this equation uses a 24-hour clock. The temperatures are given in degrees Fahrenheit. The graph is in radians, so you can enter the numbers for the hours. Figure 16-12 shows what a graph of the temperatures may look like, noting a few times and temperatures.

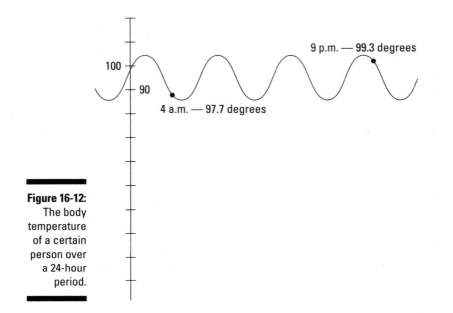

Figure 16-12:
The body
temperature
of a certain
person over
a 24-hour
period.

Making a goal

Even though people in many parts of the world play soccer year-round, certain times of the year show an increase in the sales of outdoor soccer shoes. Here's a model for the sales of pairs of shoes where N is in millions of pairs and m is the month of the year: $N(m) = 44\sin(0.524m) + 70$. From the equation, you can tell that the average number of pairs sold is 70 million, which is the vertical shift upward. That number fluctuates between 26 million and 114 million, which you find by adding and subtracting the amplitude, 44, from the average. The period of this model is $\frac{2\pi}{0.524} \approx 11.99$, or 12 months. Figure 16-13 shows a graph of this function.

A graph like the one in Figure 16-13 can help distributors and retailers with their plans for sales.

Theorizing with biorhythms

Many years ago, the public showed great interest in a person's *biorhythms*, which are the physical, emotional, and intellectual cycles that a person experiences in life. Many people even wrote books about them. Some believe that these cycles affect how a person reacts to situations in his or her life. They even go so far as to say that the positions of these curves influence major decisions of famous movie stars and politicians.

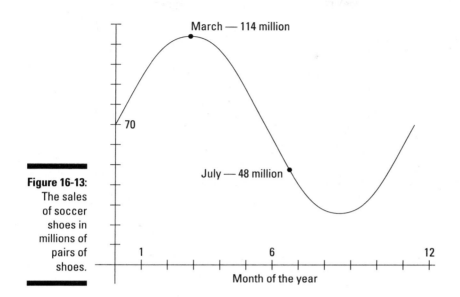

Figure 16-13:
The sales
of soccer
shoes in
millions of
pairs of
shoes.

What do biorhythms have to do with trigonometry? Everything! This bio-rhythm theory uses the sine curve. Supposedly, our life cycles start at birth and fluctuate like sine curves. The physical cycle is 23 days long, the emotional cycle is 28 days long, and the intellectual cycle is 33 days long. If you plot all these cycles on a graph, starting on the day you were born, you can see where these cycles are right now and what they'll look like in the future. Figure 16-14 shows a graph of the three biorhythm cycles.

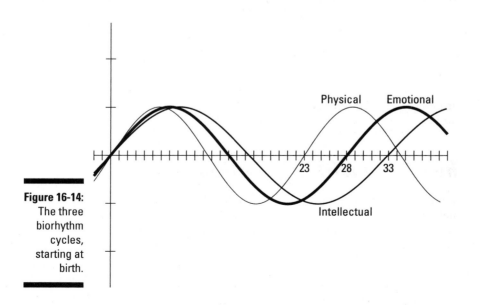

Figure 16-14:
The three
biorhythm
cycles,
starting at
birth.

In Figure 16-14, you can see how the different cycles have different periods. Imagine these sine curves going on for years and years, crossing over the *x*-axis and over one another. Figure 16-15 shows some biorhythm cycles plotted for some imaginary person for some year in the month of March.

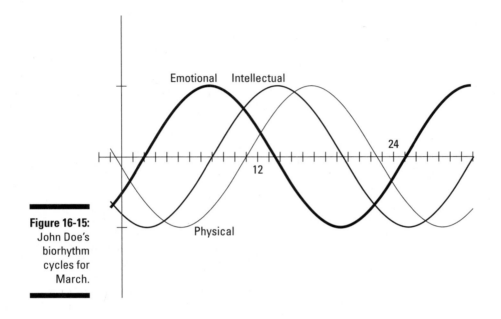

Figure 16-15:
John Doe's
biorhythm
cycles for
March.

If you believe in the biorhythm theory that says these curves exist, you can see that on the 12th of the month, all the cycles are above the *x*-axis, and on the 24th, they're all below the *x*-axis. Supposedly, when a curve is above the *x*-axis, everything is bright and sunny — a person is in good health, emotionally fine, and very smart and with it. When the curve is below the *x*-axis, he or she tends to be sick, depressed, and dull. In addition, the theory says that when the cycles cross from above to below the axis, or vice versa, those days are critical. A critical day is when upheaval and crises are possible. Such a day is a good time to stay in bed — if that's even safe. I guess there's no way to prove or disprove this theory, but it sure makes interesting use of the sine curve!

Chapter 17

Graphing Tangent and Cotangent

*T*he tangent and cotangent functions have lots of similarities. You can write both functions in terms of sine and cosine, so they share the same function values. They just don't have those function values in the same places for the *x*-values in their domain — they shift from left to right. Even though their domains (or *x*-values) are restricted, tangent and cotangent are the only trig functions with ranges (or *y*-values) that go all the way from negative infinity to positive infinity. The challenges in graphing tangent and cotangent are in dealing with the domain restrictions and *asymptotes* (dotted vertical lines used to determine the shape of a curve), as you see in this chapter.

Checking Out Tangent

The tangent function can be written as the ratio of the sine divided by the cosine: $\tan \theta = \frac{\sin \theta}{\cos \theta}$. (For more information on the tangent function, see Chapters 7 and 8.) The sine and cosine functions have values for every *x*-value, so no matter what number you put in for *x*, you'll get an answer. The only problem occurs when the cosine function is equal to 0, because a fraction can't have a 0 in the denominator. So wherever the cosine function is equal to 0, the graph of the tangent curve has asymptotes. Another interesting property comes into play with the fact that the sine and cosine are both positive in the first quadrant and negative in the third quadrant. As a result, the tangent is positive in those two quadrants and negative in the other two, because either the sine or cosine is negative, but not both.

Determining the period

The sine and cosine functions have a period of 2π, or 360 degrees, which means that after every 2π, the function pattern starts all over again. In the case of the tangent function, though, the length of the period is only π — half as long as that of sine or cosine. The tangent function repeats its pattern over and over twice as frequently as sine and cosine.

Assigning the asymptotes

An *asymptote* is a line that helps give direction to a graph. This line isn't part of the function's graph; rather, it helps determine the shape of the curve by running tangent to the curve. Asymptotes are usually indicated with dashed lines.

The asymptotes for the graph of the tangent function occur regularly, each of them π, or 180 degrees, apart. They separate each piece of the tangent curve, or each complete cycle from the next.

The equations of the tangent's asymptotes are all of the form $y = \frac{\pi}{2}(2n + 1)$, where n is an integer. Under that stipulation for n, the expression $2n + 1$ always results in an odd number. By replacing n with various integers, you get lines such as $y = -\frac{7\pi}{2}$, $y = -\frac{5\pi}{2}$, $y = -\frac{3\pi}{2}$, $y = -\frac{\pi}{2}$, $y = \frac{\pi}{2}$, $y = \frac{3\pi}{2}$, $y = \frac{5\pi}{2}$, and $y = \frac{7\pi}{2}$. The reason that asymptotes always occur at these odd multiples of $\frac{\pi}{2}$ is because those points are where the cosine function is equal to 0. As such, the domain of the tangent function includes all real numbers *except* the numbers that occur at these asymptotes.

Figure 17-1 shows what the asymptotes look like when graphed alone.

Figure 17-1 isn't all that exciting, but it does show how many times the tangent function repeats its pattern. Now take a look at Figure 17-2, which shows one cycle of the tangent function on a graph. The tangent values go infinitely high as the angle measure approaches 90 degrees. The values go infinitely low as the angle measure approaches –90 degrees.

In Figure 17-3, I put more of the tangent on a graph, asymptotes included, to give you a better idea of what's going on.

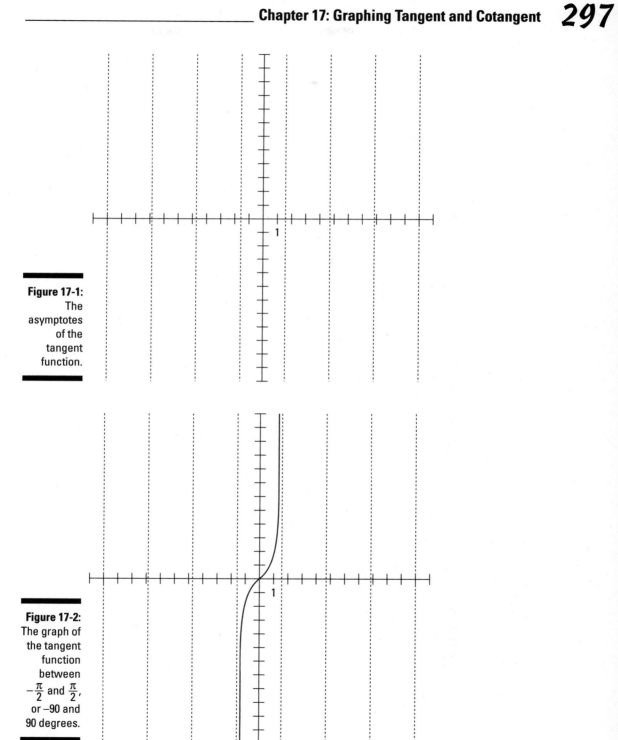

Figure 17-1:
The
asymptotes
of the
tangent
function.

Figure 17-2:
The graph of
the tangent
function
between
$-\frac{\pi}{2}$ and $\frac{\pi}{2}$,
or -90 and
90 degrees.

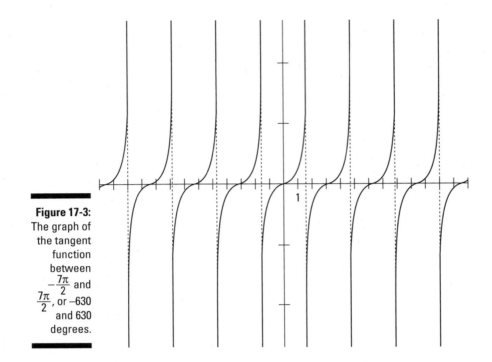

Figure 17-3:
The graph of
the tangent
function
between
$-\frac{7\pi}{2}$ and
$\frac{7\pi}{2}$, or −630
and 630
degrees.

As you can see, the tangent function repeats its values over and over. The main difference between this function and the sine and cosine functions is that the tangent has all these breaks between the cycles. As you move from left to right, the tangent appears to go up to positive infinity. It actually disappears at the top of the graph and then picks up again at the bottom, where the values come from negative infinity. Graphing calculators and other graphing utilities don't usually show the graph disappearing at the top, so it's up to you to know what's actually happening, even though the picture may not look exactly that way.

One of the peculiarities of graphing calculators is that they try to connect the tangent function to make it continuous across the screen. For this reason, you'll usually see some lines between the different parts of the curve. In a way, these lines are errors — they aren't the asymptotes, although you may be tempted to think they are. The only way to get rid of those extra lines is to turn your calculator to the *dot* mode (as opposed to the *connected* mode). Most calculators have ways to set the settings (or mode) for things such as degrees and radians, dotted graphs and connected graphs, floating decimals and fixed decimals, and so on. The changes are usually easy to do — just see your calculator's manual for specific instructions. The hard part is remembering what setting you're in.

Fiddling with the Tangent

You can alter the tangent function with multiplication, addition, and subtraction. In some cases, the effects are similar to those that occur when you alter the sine and cosine functions. Because these results aren't similar all the time, you should consider the alterations on a case-by-case basis.

Multiplying the tangent

You can multiply the tangent function by a number, but doing so doesn't affect the function the way that it affects the sine function. Multiplying the sine by a number changes its amplitude, making the function include larger and smaller values. The tangent values, however, already go from negative infinity to positive infinity. So when you multiply the entire tangent function by a number, here's what happens: If you multiply by a number bigger than 1, the function gets steeper more quickly; if you multiply by a fraction between 0 and 1, it gets flatter. And if you multiply by a negative number, the curve flips over the *x*-axis. For more on these flips (called *reflections*), go to Chapter 3. Figure 17-4 shows graphs of the basic tangent function ($y = \tan x$) and two multiples to illustrate this property.

In Figure 17-4, notice how the multiplier of 6 makes the tangent curve steeper, whereas the multiple of 0.2 makes it flatten out. Both functions still have values that go from negative infinity to positive infinity, but the rate at which they get there changes.

Multiplying the angle

Multiplying the angle variable in the tangent function has the same effect as it does with the sine and cosine functions. If the multiple is 2, as in $y = \tan 2x$, then the tangent function makes twice as many cycles in the usual amount of space. In other words, the period is $\frac{\pi}{2}$, which is the tangent's usual period, π, divided by 2. Because multiplying the angle variable of the tangent function mirrors the results of doing the same with the sine and cosine functions, I don't go into detail here — for more information, please refer to Chapter 16.

Figure 17-5 shows a few graphs to illustrate the effect of multiplying the angle variable by a number greater than 1 and then by a number between 0 and 1.

The graph of $y = \tan 3x$ doesn't show all the asymptotes, but that graph has three times as many tangent curves as usual. The graph of $y = \tan \frac{1}{2}x$ has only half as many cycles — or it takes twice as long to complete one cycle.

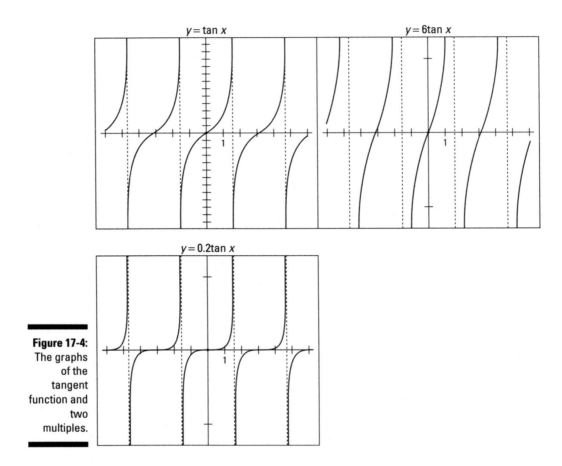

Figure 17-4:
The graphs
of the
tangent
function and
two
multiples.

Adding to tangent

Adding a number to the tangent function results in raising the curve on the graph by that amount. Likewise, subtracting a number drops the curve. Because the tangent function has values from negative infinity to positive infinity, adding to or subtracting from the function doesn't change what values the tangent has — it just changes where they happen. When you add or subtract, the *point of inflection* in the tangent curve (where the curve appears to flatten out a bit) shifts up or down. Figure 17-6 shows some graphs to illustrate this shift.

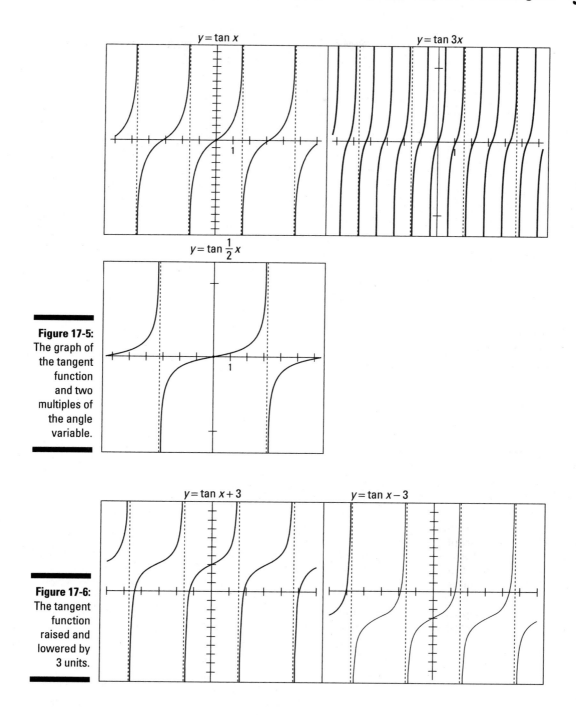

$y = \tan x$

$y = \tan 3x$

$y = \tan \frac{1}{2} x$

Figure 17-5:
The graph of
the tangent
function
and two
multiples of
the angle
variable.

$y = \tan x + 3$

$y = \tan x - 3$

Figure 17-6:
The tangent
function
raised and
lowered by
3 units.

Jacob Bernoulli

Jacob (Jacques) Bernoulli was born in Basel, Switzerland, in 1654 and died in 1705. His father owned a thriving spice business, and his mother came from an important family of bankers and local officials. Jacob's parents encouraged him to study philosophy and theology, and he graduated with a master's degree in philosophy in 1671.

While earning his degrees in philosophy, Jacob — much to the dismay of his parents — also studied mathematics and astronomy. In 1676, Jacob moved to Geneva, where he worked as a tutor. He traveled over the next few years, meeting up with leading mathematicians and scientists in Europe. He became deeply interested in astronomy and produced a work on the theory of comets, which was incorrect.

When Jacob returned to Switzerland, he taught mechanics and gave many important lectures on the mechanics of solids and liquids. Through this time, though, he never gave up on his true love for mathematics and theoretical physics.

Bernoulli made important contributions on the parallels of logic and algebra and in geometry. He's probably best known for the *lemniscate of Bernoulli,* a curve whose parametric equation is $r^2 = a^2 \cos(2\theta)$. Here's a picture of his lemniscate.

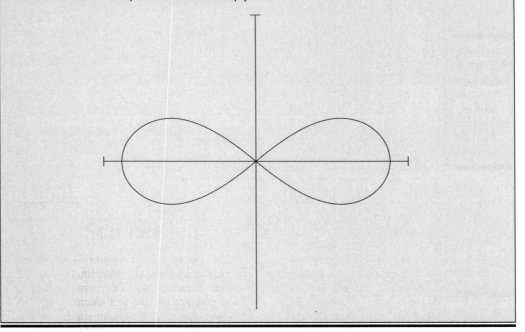

Adding or subtracting a number from the angle variable of the tangent function has the same effect as with the sine and cosine — it moves the curve to the left or right. The graph of $y = \tan(x + 1)$ shifts one unit to the left, including the asymptotes. The graph of $y = \tan(x - 1)$ moves everything to the right one unit. Figure 17-7 shows a comparison of the tangent function and the two shifted curves.

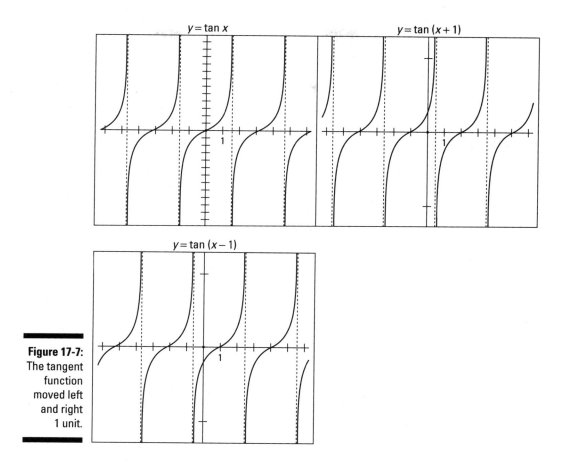

Figure 17-7:
The tangent
function
moved left
and right
1 unit.

Wherefore didst thou come, radian?

In 1873, a man named James Thomson defined and named the *radian,* the angle measure equivalent to about 57 degrees. Thomson was a mathematics professor at Queens College in Belfast, Northern Ireland. He was the brother of the famous physicist William Thomson, also known as Lord Kelvin. Although James's work seems to affect more people directly — everyone who studies or uses radian measure — his brother gained more recognition. William, Lord Kelvin, was also a mathematician who used mathematics to connect physics and electrostatics. He was the target of T.H. Huxley, an evolutionist who had some issues with mathematics and claimed that Kelvin underestimated the age of Earth.

If you have a tough time telling these graphs apart, just look for the point of inflection of the tangent curve. The point of inflection is a good reference mark when looking at all these variations.

Confronting the Cotangent

The graphs of the tangent function lay the groundwork for the graphs of the cotangent. After all, they're cofunctions and reciprocals, and have all sorts of connections. The two graphs are similar in so many ways: They both have asymptotes crossing the graph at regular intervals, go from negative infinity to positive infinity in value, and are affected by multiplying and adding. The biggest difference is in the direction the graphs are drawn. The values of the tangent function appear to *rise* as you read from left to right. The function goes upward, disappears off the graph, and then reappears down below to start all over again. The cotangent function does the opposite — it appears to *fall* when you read from left to right.

The asymptotes of the cotangent curve occur where the sine function equals 0, because $\cot \theta = \frac{\cos \theta}{\sin \theta}$. Equations of the asymptotes are of the form $y = n\pi$, where n is an integer. Some examples of the asymptotes are $y = -3\pi$, $y = -2\pi$, $y = -\pi$, $y = 0$, $y = \pi$, $y = 2\pi$, and $y = 3\pi$. (For an explanation of asymptotes, refer to the section "Assigning the asymptotes," earlier in this chapter.) Figure 17-8 shows the cotangent function graphed between -3π and 3π.

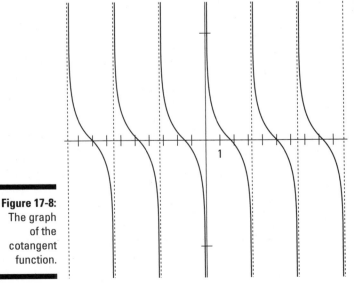

Figure 17-8:
The graph
of the
cotangent
function.

Like the other functions, cotangent repeats the same values over and over. You can apply the same types of variations to cotangent that you can to tangent (refer to the section "Fiddling with the Tangent" for the details). Figure 17-9 depicts three examples of variations: multiplying the angle variable, subtracting from the function, and adding to the angle variable.

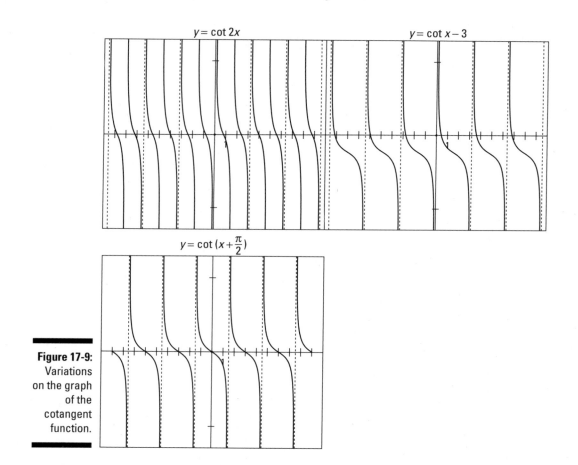

$y = \cot 2x$

$y = \cot x - 3$

$y = \cot (x + \frac{\pi}{2})$

Figure 17-9:
Variations
on the graph
of the
cotangent
function.

Chapter 18

Graphing Other Trig Functions

. . . .

. . . .

*T*he functions cosecant and secant have similarities not only because they're the reciprocals of sine and cosine, but also because their graphs look very much alike. As you see in this chapter, the easiest way to sketch the graphs of these two functions is to relate them to the graphs of their reciprocals. Doing so helps determine the *asymptotes* (where the curve approaches infinity or negative infinity), turning points, and general shape of the curves.

Seeing the Cosecant for What It Is

The cosecant function is the reciprocal of the sine function (meaning, the cosecant equals 1 divided by the sine). Even though the sine function has a domain that includes every possible number, that characteristic can't be true of its reciprocal. Whenever the sine function is equal to 0, the cosecant function doesn't exist. That fact helps determine the asymptotes you use to graph the cosecant function.

Identifying the asymptotes

The domain of the cosecant function is any number except multiples of π, because those measures are where the sine function is equal to 0. You can use this situation to identify the asymptotes by simply writing equations that use multiples of π. The asymptotes of the cosecant function are of the form $x = n\pi$, where n is some integer. Some examples of the asymptote equations are $x = -3\pi$, $x = -2\pi$, $x = -\pi$, $x = 0$, $x = \pi$, $x = 2\pi$, $x = 3\pi$, and $x = 4\pi$.

Using the sine graph

One really efficient way of graphing the cosecant function is to first make a quick sketch of the sine function. With that sketch in place, you can draw the asymptotes through the *x-intercepts* (where the curve crosses the x-axis). You can also use the maximum and minimum values on the sine function to locate the minimum and maximum points (known as *turning points*) of the cosecant function.

To graph $y = \csc x$:

    1. Sketch the graph of $y = \sin x$ from -4π to 4π, as shown in Figure 18-1.

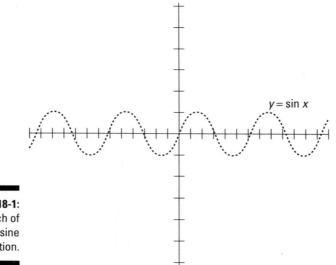

        $y = \sin x$

Figure 18-1:
A sketch of
the sine
function.

    2. Draw the vertical asymptotes through the *x*-intercepts, as Figure 18-2 shows.

    3. Draw $y = \csc x$ between the asymptotes and down to (and up to) the sine curve, as shown in Figure 18-3.

    The cosecant goes down to the top of the sine curve and up to the bottom of the sine curve.

    After using the asymptotes and reciprocal as guides to sketch the cosecant curve, you can erase those extra lines, leaving just $y = \csc x$. Figure 18-4 shows what this function looks like all on its own.

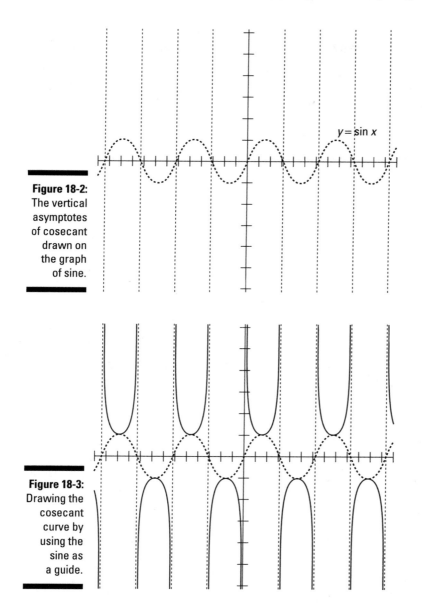

$y = \sin x$

Figure 18-2:
The vertical
asymptotes
of cosecant
drawn on
the graph
of sine.

Figure 18-3:
Drawing the
cosecant
curve by
using the
sine as
a guide.

The range of the cosecant function includes all values equal to or greater than 1 and all values equal to or less than –1. In Figure 18-4, you can see that a gap in function values lies between 1 and –1. The cosecant curve, just like all the other trig functions, keeps repeating its pattern over and over.

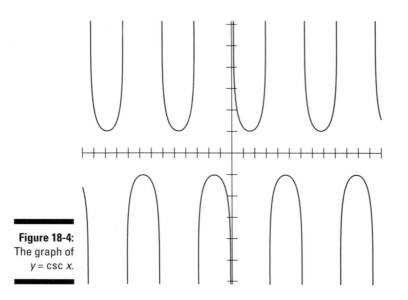

Figure 18-4:
The graph of
$y = \csc x$.

Varying the cosecant

How can you make changes to the cosecant function? This function is affected by the same multiplication, addition, and subtraction principles that affect the other functions (check out Chapter 16 for more-detailed info).

Adding or subtracting a number to or from the *cosecant function* results in slides of the graph up or down. Adding or subtracting numbers to the *angle variable* slides the graph left or right. And now I get right to it and do two slides for the price of one, sliding the graph to the left by 2 units and up by 2 units. The equation of that graph is $y = \csc (x + 2) + 2$. To find out why adding to the angle, x, moves the graph left, head on back to Chapter 3. Meanwhile, Figure 18-5 shows the graph of this equation.

Although I left out the asymptotes, you can still tell where they are — the shape of the graph is pretty clear.

Multiplying by a number changes the steepness and period of the cosecant function. If you multiply the function by 2, the curve gets steeper and has more space between its bottom and top. If you multiply the angle variable by 2, twice as much of the curve fits in the usual amount of horizontal space. Figure 18-6 shows both changes in the graph of $y = 2\csc 2x$.

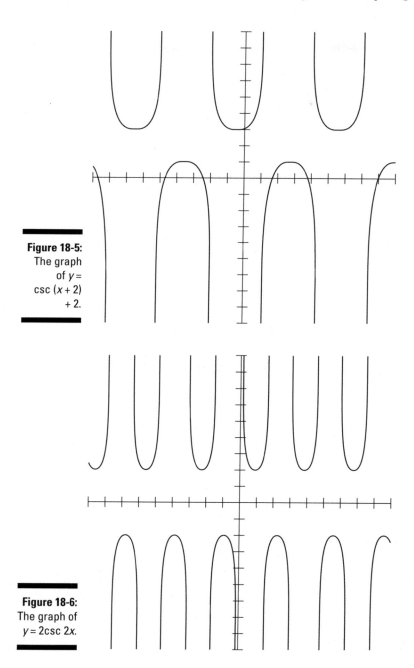

Figure 18-5:
The graph
of $y =$
csc $(x + 2)$
$+ 2$.

Figure 18-6:
The graph of
$y = 2$csc $2x$.

Unveiling the Secant

The techniques that you use to graph the secant curve parallel those that you use to graph the cosecant. First, identify the asymptotes by determining where the reciprocal of secant — cosine — is equal to 0. Then sketch in that reciprocal, and you can determine the turning points and general shape of the secant graph.

Determining the asymptotes

Because the secant equals 1 divided by the cosine, the secant function is *undefined,* or doesn't exist, whenever the cosine function is equal to 0. You can write the equations of the asymptotes by setting y equal to those values where the cosine is equal to 0, so the asymptotes are $x = -\frac{7\pi}{2}$, $x = -\frac{5\pi}{2}$, $x = -\frac{3\pi}{2}$, $x = -\frac{\pi}{2}$, $x = \frac{\pi}{2}$, $x = \frac{3\pi}{2}$, $x = \frac{5\pi}{2}$, $x = \frac{7\pi}{2}$, and so on. Another way to express the equations of all the asymptotes is to write $x = \frac{(2n+1)\pi}{2}$, where n is some integer.

Sketching the graph of secant

Using the graph of the cosine to sketch the graph of the secant function is the easiest method. Graph the cosine very lightly or with a dotted curve — the same with the asymptotes. A lot of busywork is associated with this graph, but you just have to ignore all the extra stuff and zoom in on the graph that you want. To sketch the graph of the secant function:

1. **Sketch the graph of $y = \cos x$ from -4π to 4π, as shown in Figure 18-7.**

2. **Draw the vertical asymptotes through the x-intercepts (where the curve crosses the x-axis), as Figure 18-8 shows.**

3. **Draw $y = \sec x$ between the asymptotes and down to (and up to) the cosine curve, as shown in Figure 18-9.**

   The secant goes down to the top of the cosine curve and up to the bottom of the cosine curve.

Fooling around with secant

The secant graph is different from the cosecant in several ways, but one of the most obvious ways is that this graph is symmetric about the y-axis. The secant is a mirror reflection over that axis. You can use this property to do something interesting to the graph.

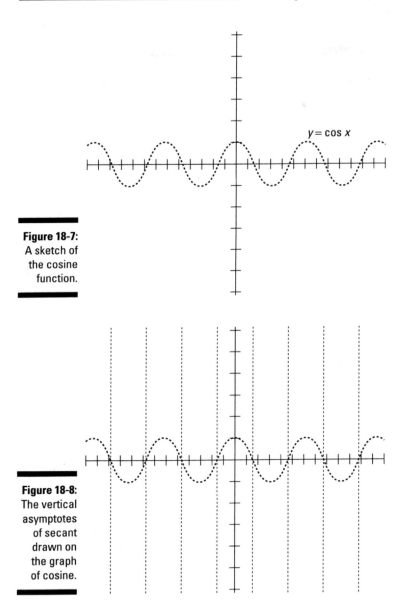

Figure 18-7:
A sketch of
the cosine
function.

Figure 18-8:
The vertical
asymptotes
of secant
drawn on
the graph
of cosine.

The usual translations and multiplications (refer to Chapter 3) affect the secant graph. If you multiply the function by $\frac{1}{6}$ and add 2π to the angle variable, as in the equation $y = \frac{1}{6}\sec(x + 2\pi)$, Figure 18-10 shows what happens.

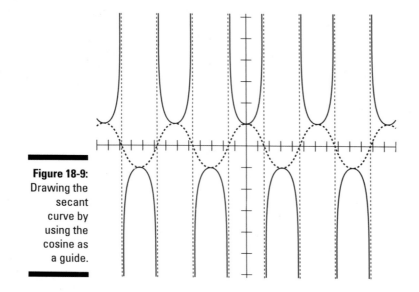

Figure 18-9:
Drawing the
secant
curve by
using the
cosine as
a guide.

Compared to $y = \sec x$, the graph in Figure 18-10 is much closer to the x-axis and seems to be flattened out between the asymptotes. These changes happen when you multiply the function by a number between 0 and 1. The turning point is still in the same place, but the y-value is much closer to 0.

Figure 18-10:
The graph of
$$y = \frac{1}{6}$$
$\sec (x + 2\pi)$.

Now you see it; now you don't

One optical illusion, called the Kanizsa Triangle, causes the eye to perceive a white equilateral triangle where none is actually drawn. Here it is:

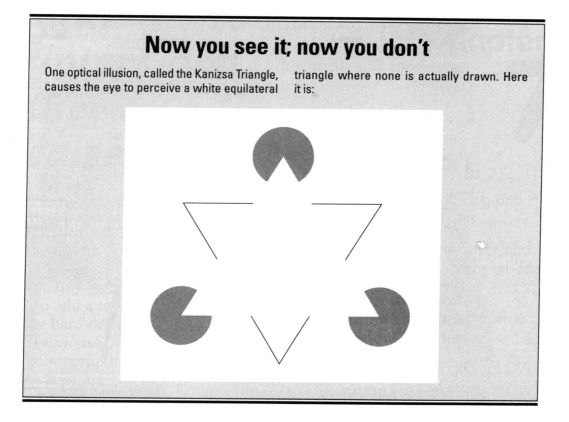

The other curiosity is that the asymptotes don't seem to be different. They aren't — and they shouldn't be. By adding 2π to the angle variable, you shift the graph 2π units to the left. The graph really has shifted, but you can't tell, because the new graph lies completely on the old one. When the shift is equal to the *period* of the function (the length of the interval that it takes for the function values to start repeating over again), then the change isn't apparent.

Laying Out the Inverse Functions

The inverse trig functions — $y = \sin^{-1} x$, $y = \cos^{-1} x$, $y = \tan^{-1} x$, $y = \cot^{-1} x$, $y = \sec^{-1} x$, and $y = \csc^{-1} x$ — are useful when solving trigonometric equations or doing applications involving trigonometry. The graphs of the inverse trig functions are rather unique; inverse sine and inverse cosine are rather abrupt and disjointed, but inverse tangent and inverse cotangent seem to go on forever, within narrow confines. Why in the world are they of any importance? For the same reason that all pictures are important — for their visual impact. Especially in the world of trig functions, remembering the general shape of a function's graph goes a long way toward helping you remember more about the function values and using them effectively.

Before diving into this section, you may want to go back and review the material in Chapter 3 on inverse functions if you need to reacquaint yourself with the domains and ranges of these functions and their respective values.

Graphing inverse sine and cosine

The first two graphs sort of go together — they have a common characteristic. The input values for both $y = \sin^{-1} x$ and $y = \cos^{-1} x$ are all the numbers from –1 to 1, including those numbers. The inputs are restricted to those values because they're the *output* values of the sine and cosine.

The output, or range, values for these two inverse functions are different. The range of $y = \sin^{-1} x$ consists of angles in the first and fourth quadrants. In radians, the range is $-\frac{\pi}{2}$ to $\frac{\pi}{2}$; in approximate decimal values, the range is –1.571 to 1.571. The range of $y = \cos^{-1} x$, on the other hand, consists of angles in the first and second quadrants, or angles from 0 to π. In approximate decimal values, that range is 0 to 3.142.

Figure 18-11 shows what the graphs of inverse sine and cosine look like.

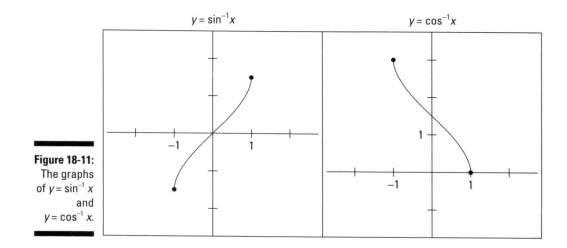

Figure 18-11:
The graphs
of $y = \sin^{-1} x$
and
$y = \cos^{-1} x$.

The points indicated on the graphs are at $x = -1$ and $x = 1$. These points are the extreme values of the inputs. The y-values represent the angle measures. If you want to find a point on either graph, just find some number between -1 and 1, and find the place on the graph corresponding to that x-value.

Taking on inverse tangent and cotangent

The tangent and cotangent functions have restricted inputs — certain angles don't jibe with them. But their outputs go through all the real numbers. If you switch those two groups of numbers to fit the inverses of tangent and cotangent, you can say that the *inputs* go through all the real numbers, and the *outputs* are restricted. The graphs of these two inverse functions are quite interesting because they involve two horizontal asymptotes. The asymptotes help with the shapes of the curves and emphasize the fact that some angles won't work with the functions.

The two horizontal asymptotes for the inverse tangent function are $y = -\frac{\pi}{2}$ and $y = \frac{\pi}{2}$, because the tangent function doesn't exist for those two angle measures. The tangent function isn't defined wherever the cosine is equal to 0. If you need to review the tangent function, go to Chapters 7 and 8. The graph of the inverse tangent has x-values from negative infinity to positive infinity, with all y-values between those two asymptotes.

The two horizontal asymptotes for the inverse cotangent function are $y = 0$ and $y = \pi$. As with the inverse tangent, the inverse cotangent function goes from negative infinity to positive infinity between the asymptotes. Check out both graphs in Figure 18-12.

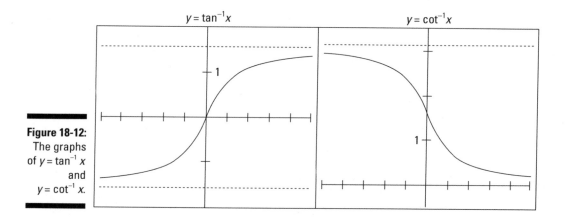

Figure 18-12:
The graphs
of $y = \tan^{-1} x$
and
$y = \cot^{-1} x$.

The main differences between these two graphs is that the inverse tangent curve *rises* as you go from left to right, and the inverse cotangent *falls* as you go from left to right. Also, the horizontal asymptotes for inverse tangent capture the angle measures for the first and fourth quadrants; the horizontal asymptotes for inverse cotangent capture the first and second quadrants. The measures between these asymptotes are, of course, consistent with the ranges of the two inverse functions.

Crafting inverse secant and cosecant

The graphs of the inverse secant and inverse cosecant will take a little explaining. First, keep in mind that the secant and cosecant functions don't have any output values (*y*-values) between –1 and 1, so a wide-open space plops itself in the middle of their graphs. This idea translates into a wide-open space between the *x*-values –1 and 1 in the graphs of their inverses. Also, the graphs of secant and cosecant go infinitely high and infinitely low along the *y*-axis. So the graphs of the inverses have a horizontal asymptote. All this talk probably seems like nonsense, so take a look at Figure 18-13, which shows the graphs.

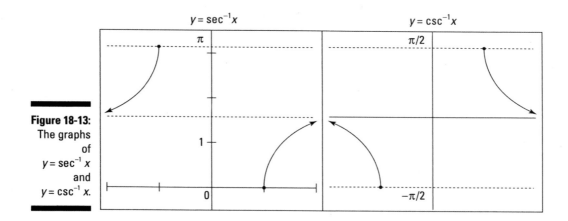

Figure 18-13: The graphs of $y = \sec^{-1} x$ and $y = \csc^{-1} x$.

The graph of $y = \sec^{-1} x$ lies between 0 and π on the *y*-axis. All the output values are in the first and second quadrants. But a horizontal asymptote runs through the graph: the line $y = \frac{\pi}{2}$. The secant isn't defined at $\frac{\pi}{2}$, so its inverse won't have an output value there. The graph of the inverse secant goes from the point (1,0) and moves upward, staying below the horizontal asymptote as the *x*-values go to positive infinity. It also comes from negative infinity along the *x*-axis above the horizontal asymptote, moving upward to the point (–1,π).

The graph of $y = \csc^{-1} x$ lies between $-\frac{\pi}{2}$ and $\frac{\pi}{2}$, with a horizontal asymptote of $y = 0$. (The cosecant isn't defined at $x = 0$, so its inverse doesn't have an output value there.) The graph of inverse cosecant covers angle measures from the first and fourth quadrants. On the right, the graph goes from the point $\left(1, \frac{\pi}{2}\right)$ down toward the horizontal asymptote as the x-values go to positive infinity. On the left, the graph's x-values come from negative infinity, where they're just below the asymptote, and move down to the point $\left(-1, -\frac{\pi}{2}\right)$.

Chapter 19

Topping Off Trig Graphs

- -

In This Chapter

▶ Identifying the graph from the trig equation

▶ Combining functions to fit real-life applications

▶ Comparing graphs to everyday scenarios

- -

The graphs of the trigonometric functions can take on many variations in their shapes and sizes. As wonderful as they are just by themselves, they're even better and more useful when you adjust them to fit a particular situation. In Chapters 16, 17, and 18, I show you how to make the trig functions slide about by moving them up, down, left, and right. I also show you how to make them steeper and flatter. In this chapter, I complete the trig story with additional transformations, as well as the even-more-exciting possibilities that occur when you combine graphs. I start off with a basic template for a trig function and progress from that point.

The Basics of Trig Equations

You can identify all the different transformations that you can perform on a trig function from a certain form of the function's equation. First, check out the general equation and then consider some examples of what the specific equations may look like.

The general form for a trig equation is $y = Af\left[B(x+C)\right]+D$, where

▶ f represents the trig function

▶ A represents the *amplitude,* or steepness

- +A means the graph is oriented as usual

- –A means that the graph is flipped over a horizontal line

▶ B determines the *period* of the graph (the length of the interval needed for the graph of the function to start repeating itself)

✔ *C* determines a shift to the left or right

✔ *D* determines a shift up or down

Some examples of trig functions using this format are

$$y = 3\sin\left[2\left(x - \tfrac{\pi}{4}\right)\right] + 1$$

$$y = -2\cos\left[\tfrac{1}{3}(x + \pi)\right] - 3$$

$$y = -\tan 2x + \pi$$

Each of the numbers changes the basic graph in a particular way. I graph the preceding three equations in the section "Graphing with the General Form," later in this chapter, but first I explain what happens when the first number, *A*, is negative.

Flipping over a horizontal line

When you multiply the trig function by a negative number, then all the output values are reversed. The positive values become negative, and the negative values become positive. The effect that this operation has on the graph is that it appears to have a reflection or flip over a horizontal line. For example, Figure 19-1 shows the graph of *y* = –sin *x* compared to *y* = sin *x*:

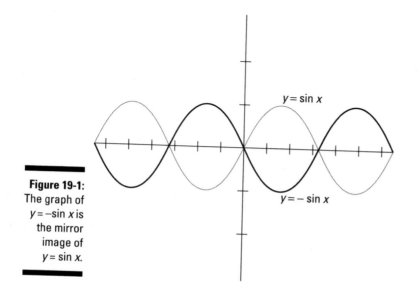

Figure 19-1:
The graph of
y = –sin *x* is
the mirror
image of
y = sin *x*.

See how the two graphs compare? The original graph appears to be flipped over the *x*-axis.

Interpreting the equation

The different letters of the general equation for a trig function each have a purpose. Here's a more-detailed explanation of each part.

A is for amplitude

The letter *A* represents the amplitude of the sine or cosine function, and it affects the steepness or flatness of the graphs of any of the trig functions. If the absolute value (ignore the + or − sign) of *A* is some number greater than 1, then the graph is steeper than usual. If the absolute value of *A* is between 0 and 1, then the graph is flatter. The higher the number, the steeper the curve. The closer the number is to 0, the flatter the curve.

B is for becoming (The period)

The multiplier *B* affects the length of the graph's *period,* or how far it goes along the *x*-axis. The sine, cosine, cosecant, and secant all normally have a period of 2π. The tangent and cotangent have a period of π. If you divide the normal period of the function by the value of *B*, then you get the length of the new, adjusted period. Another way to put it: *B* tells you how many complete cycles the curve will make in the space that usually has only one. If *B* is 2, then the graph has two complete cycles where there's usually one.

C is for cruisin' left or right

The value of *C* changes the graph by moving the whole curve to the left or right of where it usually is. If you subtract *C*, the graph moves *C* units to the right. If you add *C*, it moves *C* units to the left.

D is for distancing yourself

The value of *D* tells how far up or down the graph moves from its original position. A positive *D* moves the graph up, and a negative *D* moves it down. The value of *D* also represents the average or middle value of the sine and cosine curves and the middle of the open space of the secant and cosecant curves.

Graphing with the General Form

Now is the time to put all your knowledge to work and do some serious graphing. In the general equation for a trig function, $y = Af\left[B(x+C)\right] + D$, the letters *A*, *B*, *C*, and *D* all represent values, but they have to be in those exact places, and the equation has to be in that exact form. You need to factor, multiply, or manipulate the equation in other ways to get it in the general form if you want to use these values to figure out what the graph looks like.

More likely than not, you'd draw the graphs in this section with the help of a graphing calculator. Technology really has come a long way in helping math students and others who use trigonometry. The only problem with graphing calculators is that entering these complicated functions correctly is often a real challenge. You can't use brackets or braces to help keep the groupings straight. You're stuck with parentheses, which can get messy when you're dealing with a lot of them. The main reason I provide the examples in this section is so you know what to expect. When you know how all these variations work, you're able to recognize when you have an error in your graphing calculator work and avoid that age-old saying, "Garbage in, garbage out."

The first example involves graphing $y = 3\sin\left[2\left(x - \frac{\pi}{4}\right)\right] + 1$.

1. **Determine the amplitude of the curve.**

   The 3 represents the A, which is the amplitude of a sine curve. The function will stretch 3 units above and below the middle. The 3 is positive, so the curve doesn't flip or reflect over a horizontal line.

2. **Find the period of the function.**

   The 2 represents the B, which means that the curve makes two complete cycles in the amount of space where it usually has only one. Because the normal period of the sine function is 2π units, in this function the period is $\frac{2\pi}{2}$, or π units long.

3. **Determine the shift left or right.**

   The $-\frac{\pi}{4}$ represents the value of C. Because C is negative, the shift is $\frac{\pi}{4}$ units to the *right*.

4. **Find the shift up or down.**

   The last number, 1, is the value of D, which is a shift upward by 1 unit.

5. **Input all the values to graph the equation, as shown in Figure 19-2.**

In Figure 19-2, I drew the line $y = 1$ to show the middle, so you can see the result of the vertical shift. The graph goes from about -2π to 2π on the x-axis, where you'd normally expect to find two complete cycles. Instead, the graph has four.

The next graph has a flip over a horizontal line. The curve doesn't flip over the x-axis because the graph is dropped down by 3 units. Instead, the flip is over the horizontal line $y = -3$. Without further ado, here's how to sketch the graph of $y = -2\cos\left[\frac{1}{3}(x + \pi)\right] - 3$.

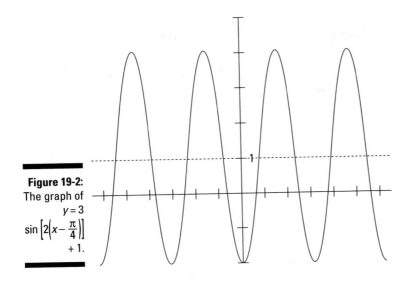

Figure 19-2:
The graph of
$y = 3$
$\sin\left[2\left(x - \frac{\pi}{4}\right)\right]$
$+ 1.$

1. **Determine the amplitude.**

   The –2 in front tells you two things. First, the amplitude is 2, or the curve is twice as high as usual. The negative sign tells you that the whole cosine curve is flipped over a horizontal line. Where the curve usually goes up, it goes down, and vice versa.

2. **Find the period.**

   The multiplier of $\frac{1}{3}$ spreads the curve out quite a bit — only one-third as much curve is in the same amount of space as a 2π period (the period of the basic cosine function) usually has. In fact, the new period is

   $2\pi \div \frac{1}{3} = 2\pi \cdot 3 = 6\pi.$

3. **Determine the shift left or right.**

   The value of C is π, which is a little more than 3 units. The graph moves 3 units to the *left*, because C is positive.

4. **Find the vertical shift.**

   D represents the number –3, so the whole graph shifts down 3 units.

5. **Now use all these values to graph the curve from -2π to 2π, as Figure 19-3 shows.**

The graph from −2π to 2π   The graph from −4π to 4π

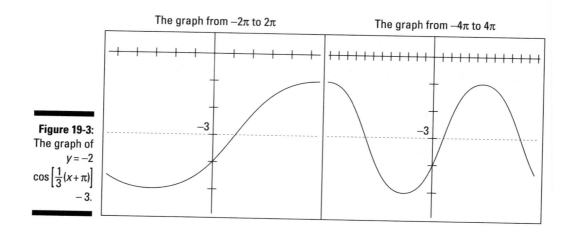

Figure 19-3:
The graph of
$y = -2$
$\cos\left[\frac{1}{3}(x+\pi)\right]$
-3.

The seed of life

A *compass* is an instrument that you can use to draw a circle. All you do is place the sharp, pointed end at the center of the circle and drag the pencil end around. One of the basic constructions that students can produce is a set of circles that intersect with one another in an interesting pattern. When constructed correctly, those circles form what's called the *seed of life,* an arrangement that has ancient Egyptian beginnings. This pattern is also a part of 13th-century Italian art. Here's what it looks like.

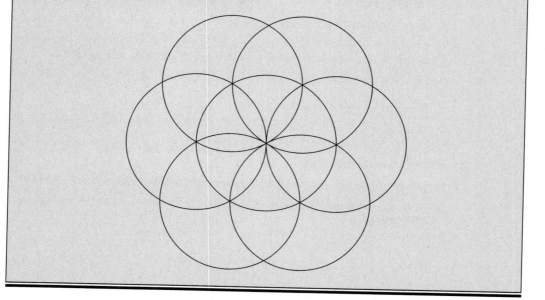

As you can see from the graphs in Figure 19-3, just graphing from -2π to 2π doesn't show a complete cycle. The period is 6π, so the graph needs more space. The two graphs look a little different — the one on the right looks steeper, because the scales on the x-axes are different. Both graphs are of the same function. The dotted line is $y = -3$, which is the middle or average of the graph.

This last graph shows that you don't always have to have a value for one of the letters in the general form. Well, actually, each of the letters always has a value, but when that value is 1 or 0, it doesn't show up. Just know that the part of the equation with the value of 1 or 0 doesn't change the basic graph. In the graph of $y = -\tan 2x + \pi$, you find two situations where the original graph doesn't change.

1. **Determine the steepness.**

   With the tangent function, I don't refer to the multiplier as *amplitude* because the tangent curve doesn't have a highest or lowest point, as the sine and cosine curves do. Any multiplier A affects the steepness. In this case, that steepness doesn't change, because the A is essentially a 1. Because the 1 is negative, the graph flips over a horizontal line.

2. **Find the period of the function.**

   The multiplier of 2 makes the period of this tangent curve equal to $\frac{\pi}{2}$, because the normal period of the tangent function is π, and you have to divide by 2 here. The graph makes twice as many cycles in the usual amount of space.

3. **Determine the shift left or right.**

   Here's another case where the graph doesn't change. The equation has no number in place of C — that value is actually 0. So the graph doesn't shift left or right.

4. **Find the vertical shift.**

   The number π is the D value. That number is positive, so the graph shifts up π units.

5. **Graph the function from -2π to 2π, as Figure 19-4 shows.**

You can see that the graph in Figure 19-4 shows eight complete cycles of the tangent function. The reason it shows so many cycles is because the graph goes from -2π to 2π, and each cycle is only $\frac{\pi}{2}$ in length. The dotted line shows the horizontal shift of π; I left the vertical asymptotes out.

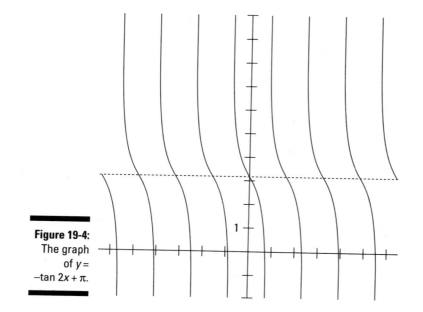

Figure 19-4:
The graph
of $y =$
$-\tan 2x + \pi$.

Adding and Subtracting Functions

Just when you thought this book couldn't get any better, I add yet another twist to the trigonometry picture. You can model many applications in physics and the cycles of nature with curves that you create by adding or subtracting two trig functions together or by adding a trig function and some algebraic function. When you add functions together, you can obtain the graph of this sum by taking each x-value from each function and finding the sum of the y-values that correspond to those x-values. Then you can plot the points with the x-values that you used and the y-values that you found. I show you a couple of examples, and because they get too messy very quickly, I bail and use a graphing calculator.

The function $y = x + \sin x$ is the sum of the sine function $y = \sin x$ and the algebraic function $y = x$. The algebraic function $y = x$ is a line that cuts diagonally through the third and first quadrants. The sine has y-values that go from -1 to 1, over and over again. Table 19-1 shows some of the separate functions' values and then the sum of those values.

Table 19-1				The *y*-Values of $y = x + \sin x$						
x	-2π	$-\pi$	-2	-1	$-\frac{\pi}{2}$	0	$\frac{\pi}{2}$	1	2	π	2π
sin x	0	0	-0.909	-0.841	-1	0	1	0.841	0.909	0	0
x + sin x	-2π	$-\pi$	-2.909	-1.841	-2.571	0	2.571	1.841	2.909	π	2π

You can get a better idea of how this addition works by looking at the graph. I used a graphing calculator to graph $y = x + \sin x$ from -2π to 2π. And then, to give you an even better picture of what's going on (and because I think this curve is neat), I graphed it from -4π to 4π in Figure 19-5.

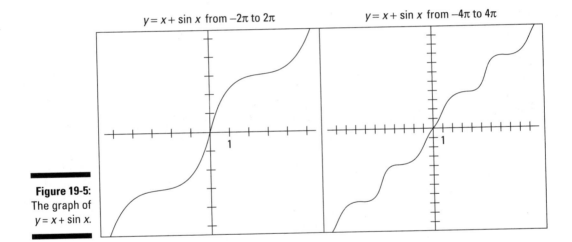

$y = x + \sin x$ from -2π to 2π $y = x + \sin x$ from -4π to 4π

Figure 19-5:
The graph of
$y = x + \sin x.$

The next example shows what can happen when you subtract one trig function from another. Of course, I had to experiment with all sorts of different combinations of functions to make this graph come out especially interesting. You should try your hand at it, too. Here's my contribution, the graph of $y = 2\sin x - \cos 3x$. I subtracted the function values of each, one from the other, to produce the very nice curve in Figure 19-6, which I drew from -4π to 4π. Is it a heartbeat or a pretty design?

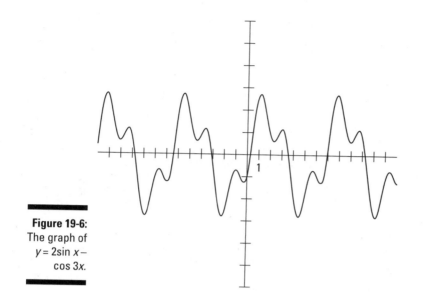

Figure 19-6:
The graph of
$y = 2\sin x - \cos 3x$.

Applying Yourself to the Task

The graphs of some of the trig functions that you can create by altering the functions or combining them are fun to look at. They may even be useful when you're preparing a special border or other artwork. But the practical uses of these graphs are what you consider in this section. A cardiologist looks at a graph of the heart's function and detects whether it's beating properly. The graphs of earthquake activity are of special interest to those hoping to predict the next one — with enough time to warn everyone.

Measuring the tide

Along the coast, the tides are of particular interest. The tides are affected by the gravitational pull of both the moon and the sun. The high tides and low tides follow a periodic pattern that you can model with the sine function. On a particular winter day, the high tide in Boston, Massachusetts, occurred at midnight. To determine the height of the water in the harbor, use the equation $H(t) = 4.8\sin\frac{\pi}{6}(t+3) + 5.1$, where t represents the number of hours since midnight.

1. **Input the value of t into the equation.**

   At midnight, the value of t is 0. Putting 0 in for t in the equation gives you

$$H(0) = 4.8\sin\frac{\pi}{6}(3) + 5.1$$

$$= 4.8\sin\frac{\pi}{2} + 5.1$$

$$= 4.8(1) + 5.1 = 9.9$$

The greatest value of sine occurs at $\frac{\pi}{2}$, so it makes sense that high tide would be when the formula uses the sine of that value.

2. **Determine the altitude.**

   The multiplier of 4.8 is the amplitude — how far above and below the middle value that the graph goes. The tides go 4.8 feet above and below the average amount on this particular day. The number added on at the end, 5.1, is the average height for the tides. So the tide goes up to 9.9 feet and down to 0.3 feet — wading depth.

3. **Find the period of the function.**

   The multiplier of $\frac{\pi}{6}$ affects the period. The period of the sine function is usually 2π. Divide 2π by $\frac{\pi}{6}$, and you get $2\pi \div \frac{\pi}{6} = 2\pi \cdot \frac{6}{\pi} = 12$. The period is 12 hours, so you know that the tides go through their entire cycle in 12 hours. The 3 added to the t is a shift horizontally; that number determines what times of day the high tide and low tide occur. Figure 19-7 shows a graph of this function and the different stages of the tide at different times. By looking at the graph, you can plan your sailing and clam-digging activities.

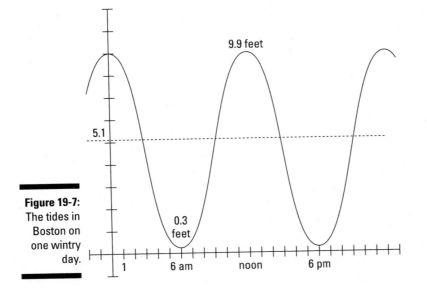

Figure 19-7:
The tides in
Boston on
one wintry
day.

Tracking the deer population

The graph in this section shows the population of a herd of deer, starting at the first of April and ending at the next April. New deer are born in the spring, so an increase in the herd size is expected. Predators take care of the weak deer — both young and old. And then you have to consider the weather; winter can be very hard on the population. Look at the graph in Figure 19-8 and see if it demonstrates what you'd expect. This cycle is the result of finding the sum of two different sine functions: $D(m) = 400 + 40\sin(0.524m) + 20\sin(1.047m)$ represents the population of the herd, where m is the number of months since April.

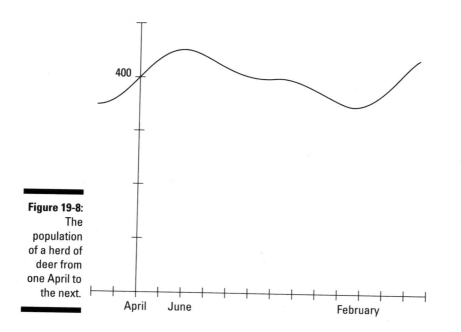

Figure 19-8:
The population of a herd of deer from one April to the next.

The herd experiences a high of about 450 deer in June and a low of about 350 deer in February. This model shows a herd that stays pretty close to being the same size year after year.

Measuring the movement of an object on a spring

In this section, a trig function proves useful in a model for an object attached to a spring. The same pattern doesn't occur over and over, as it does in the previous sections. But this is a great example of a trig function at work.

The equation $H(t) = 3(0.7)^t \cos 5t + 4$ represents the height of an object attached to a spring, where t is the amount of time that has passed — usually in seconds. The equation has a trig function multiplied by an exponential function. When you first release the spring, the object hits a height of about 7 feet. It jumps up and down, finally settling in at about 4 feet high, as shown in Figure 19-9.

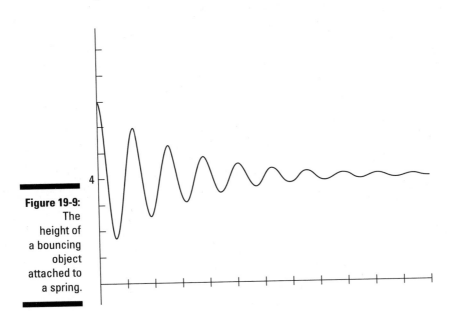

Figure 19-9: The height of a bouncing object attached to a spring.

You can probably come up with a similar model to show how a bungee jumper goes up and down. So you see, the trig functions have all sorts of applications — many of them very useful.

Part VI
The Part of Tens

The 5th Wave   By Rich Tennant

In this part . . .

In this part, I lay it all out on the line by providing some basic and some rather obscure trig identities that you can quickly turn to in a time of need. I also tell you everything you've ever wanted to know about the area of a triangle — but were afraid to ask. Lastly, if you want to try your hand at weather forecasting, you can predict the temperatures in some selected cities by using some handy sine-curve equations.

Chapter 20

Ten Basic Identities . . . Plus One

A big advantage of trig expressions and equations is that you can adjust them in so many ways to suit your needs. The basic identities that I list in this chapter are the ones people use most frequently (and remember most often).

Reciprocal Identities

Take a look at the first reciprocal identity and its counterpart:

$$\sin \theta = \frac{1}{\csc \theta} \text{ and } \csc \theta = \frac{1}{\sin \theta}$$

Secant, cosecant, and cotangent are technically the three reciprocal functions, but you can write identities to show their reciprocals, too. Next are the second reciprocal identity and its counterpart.

$$\cos \theta = \frac{1}{\sec \theta} \text{ and } \sec \theta = \frac{1}{\cos \theta}$$

The tangent and its reciprocal at least have names that sound alike. The other two basic functions and their reciprocals (see the preceding equations) don't seem to have any rhyme or reason to their pairing.

$$\tan \theta = \frac{1}{\cot \theta} \text{ and } \cot \theta = \frac{1}{\tan \theta}$$

Ratio Identities

Both of the ratio identities involve fractions with sine and cosine.

$$\tan \theta = \frac{\sin \theta}{\cos \theta}$$

$$\cot \theta = \frac{\cos \theta}{\sin \theta}$$

 Here's a helpful way to remember which ratio identity has the sine in the numerator: Tangent and sine have beginning letters that are very close in the alphabet, and cotangent and cosine have the same beginning letters. This train of thought helped me out in high school when I first saw these identities.

Pythagorean Identities

The first Pythagorean identity uses your good friends sine and cosine and is probably one of the most frequently used identities.

$$\sin^2 \theta + \cos^2 \theta = 1$$

This second Pythagorean identity comes from the first Pythagorean identity; simply divide each term in that identity by the square of the cosine function and simplify.

$$\tan^2 \theta + 1 = \sec^2 \theta$$

Last but certainly not least, you get the third Pythagorean identity from the first one by dividing each of the terms in that identity by the square of the sine.

$$1 + \cot^2 \theta = \csc^2 \theta$$

Opposite-Angle Identities

This opposite-angle identity allows you to change to a positive angle when using the sine function:

$$\sin(-\theta) = -\sin \theta$$

This second opposite-angle identity may seem a little odd — the fact that the cosine of a negative angle has the same value as the cosine of the corresponding positive angle is always a little puzzling at first.

$$\cos(-\theta) = \cos\theta$$

The opposite-angle identity for the tangent follows the same pattern as that of the sine:

$$\tan(-\theta) = -\tan\theta$$

Chapter 21

Ten Not-So-Basic Identities

*I*n Chapters 10 and 11, I cover the most frequently used identities at great length. Here are ten identities that don't appear in those chapters, because you won't use them all that often. A few are rather obscure. These identities don't lend themselves to memorization very well — you'll be better off just looking them up if you need them.

Product-to-Sum Identities

The product-to-sum identities look very much alike. You have to pay close attention to the subtle differences so that you can apply them correctly. Even though the product looks nice and compact, it's not always as easy to deal with in calculus computations — the sum or difference of two different angles is preferred.

The first identity has two angles, *A* and *B*. When you multiply the sine of one angle times the cosine of the other angle, you end up with one-half the sum of a sum identity and a difference identity. Whew!

$$\sin A \cos B = \frac{1}{2}\Big[\sin(A+B) + \sin(A-B)\Big]$$

This time, multiply the sines of both angles together, and the result equals one-half the difference between a sum identity and a difference identity:

$$\sin A \sin B = \frac{1}{2}\Big[\cos(A-B) - \cos(A+B)\Big]$$

This identity has a *mix-and-match* feel to it. Two different angles and two different functions are used. There seems to be something for everyone.

$$\cos A \sin B = \frac{1}{2}\big[\sin(A+B) - \sin(A-B)\big]$$

The last product-to-sum identity uses the cosines of two angles:

$$\cos A \cos B = \frac{1}{2}\big[\cos(A-B) + \cos(A+B)\big]$$

Sum-to-Product Identities

The sum to product identities are useful for modeling what happens with sound frequencies. Think of two different tones represented by sine curves. Add them together, and they beat against each other with a warble — how much depends on their individual frequencies. The identities give a function modeling what's happening.

The first identity takes two different angles, *A* and *B*, and adds their sines together. The result: twice the product of the sine and cosine of two new angles that are created by a fraction. See for yourself:

$$\sin A + \sin B = 2\sin\left(\frac{A+B}{2}\right)\cos\left(\frac{A-B}{2}\right)$$

You can technically call this next identity a *difference*-to-product identity, although math gurus usually classify it with the sum-to-product identities. Of course, you can consider the difference to be a sum if you call it the sum of a sine and the opposite of another sine.

$$\sin A - \sin B = 2\cos\left(\frac{A+B}{2}\right)\sin\left(\frac{A-B}{2}\right)$$

This next identity involves the sum of the cosines of two angles.

$$\cos A + \cos B = 2\cos\left(\frac{A+B}{2}\right)\cos\left(\frac{A-B}{2}\right)$$

As you probably expect, the last sum-to-product identity has the difference of the cosines of two angles.

$$\cos A - \cos B = -2\sin\left(\frac{A+B}{2}\right)\sin\left(\frac{A-B}{2}\right)$$

Reduction Formula

The reduction formula reduces two trig functions into one. It's useful when studying the force of a spring, the position of a swinging pendulum, or the current in an electrical circuit. This formula takes the sum of two different functions with the common input x and changes the sum to a single function where the multiplier of the sine is the amplitude and the phase shift is θ. The a and b are the coordinates of some point on the terminal side of θ when it's in standard position:

$$a \sin x + b \cos x = \sqrt{a^2 + b^2} \sin(x + \theta)$$

To get the sine and cosine of x, you can use simplified versions:

$$\sin x = \frac{b}{\sqrt{a^2 + b^2}} \text{ and } \cos x = \frac{a}{\sqrt{a^2 + b^2}}.$$

Mollweide's Equations

Karl Mollweide was an astronomy teacher. His work in astronomy and mathematics led to his discovery of identities that can take the measures of the sides of a triangle and relate them to an expression involving trig functions. Mollweide's equations involve all six parts of a triangle: the three angles, A, B, and C, and the three corresponding sides opposite those angles, a, b, and c.

$$\frac{a + b}{c} = \frac{\cos \frac{1}{2}(A - B)}{\sin \frac{1}{2} C} \text{ or } \frac{a - b}{c} = \frac{\sin \frac{1}{2}(A - B)}{\cos \frac{1}{2} C}$$

Chapter 22

Ten Ways to Find the Area of a Triangle

The triangle is one of the more-familiar geometric figures, so people have studied it and restudied it ad nauseum. Give me some information about a triangle, and I can tell you its area. Why might you want the area? Because it has many practical applications, such as finding the area of a yard so you can buy enough fertilizer or figuring out just how big that sail for your sailboat has to be.

Several of the formulas in this chapter are based on a triangle with its vertices labeled *A, B,* and *C* and the sides opposite those vertices labeled *a, b,* and *c,* respectively. Figure 22-1 shows a sketch of a triangle labeled this way.

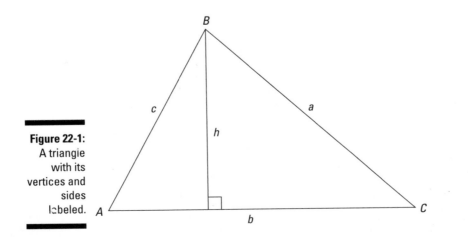

Figure 22-1:
A triangle with its vertices and sides labeled.

Using the Base and Height

The first formula for finding the area of a triangle uses the base and the height. The *base* is any side of the triangle, and the *height* (also called the *altitude*) is the distance from the vertex opposite that chosen side drawn perpendicular to it. To use this formula, you must know for a fact that the altitude and side meet with a right angle. In Figure 22-1, the height, *h,* is drawn from the vertex *B* to the side *b*. The length of side *b* is the base, represented by the letter *b* in the formula.

$$Area = \frac{1}{2} bh$$

Now forget Figure 22-1 for a second. The next formula looks just like the first one except for one letter. For this formula, *a* and *b* represent two perpendicular legs of a right triangle. You don't have to draw in an altitude on a right triangle — you can just use those two perpendicular sides.

$$Area = \frac{1}{2} ab$$

Acquainting Yourself with Heron

Heron's Formula goes like this:

$$Area = \sqrt{s(s-a)(s-b)(s-c)}$$

All you need to find the area with this method are the measures of the three sides. The *s* represents the *semiperimeter,* which is half of the perimeter. Just add up the lengths of the sides, take half of that sum, and insert it into the formula.

Applying Trig (The Whole Point of This Book)

You have all sorts of ways to find the area of a triangle — you can purely use measures of length, or you can get into the angle business. These next formulas may look like the law of sines or the law of cosines, but take note that these equations are for finding *area,* and those laws are for finding missing measures of the triangles. Of course, the law of sines or cosines can come in handy to fill in missing details.

Side-angle-side formulas

These formulas are called the *SAS* or *side-angle-side* formulas, because you can find the area if you have the measures of two sides and the angle between them. They're easy to remember, because you choose two letters for the sides and then use the angle whose letter you haven't used yet. Unless the angle you're dealing with has a nice, simple value for the sine, you need a calculator or trig table to compute the area.

$$Area = \frac{1}{2}\,ab\sin C$$

$$Area = \frac{1}{2}\,bc\sin A$$

$$Area = \frac{1}{2}\,ac\sin B$$

Angle-side-angle formulas

These three formulas work when you have two angles and the side between them, so they're called the *ASA* or *angle-side-angle* formulas. They're also sort of easy to remember, because the side squared on the top has its opposite angle on the bottom. Whatever works. . . . Notice that they all use the sine function — more trigonometry.

$$Area = \frac{a^2 \sin B \sin C}{2\sin A}$$

$$Area = \frac{b^2 \sin A \sin C}{2\sin B}$$

$$Area = \frac{c^2 \sin A \sin B}{2\sin C}$$

Going Back to Algebra: Using Determinants

The following formula uses *determinants* (rectangular arrangements of numbers). You evaluate determinants by multiplying the numbers diagonally in one direction, adding the products, and then subtracting a multiplication in the other direction. Using a graphing calculator to evaluate these determinants is really much more practical, though, because the computations can get a bit messy.

To write a determinant, list all the *x*-coordinates in the first column and their corresponding *y*-coordinates in the second column (see Figure 22-2 for these coordinates on a graph). Then write all 1s in the third column. You have to list the points from top to bottom in counterclockwise order, but you can start with any point.

$$Area = \begin{vmatrix} x_1 & y_1 & 1 \\ x_2 & y_2 & 1 \\ x_3 & y_3 & 1 \end{vmatrix}$$

Now just haul out your calculator to evaluate the determinant, or do the multiplying yourself. If you're multiplying on your own, rewrite the first two columns to the right of the determinant to make the multiplication easier:

$$Area = \begin{vmatrix} x_1 & y_1 & 1 \\ x_2 & y_2 & 1 \\ x_3 & y_3 & 1 \end{vmatrix} \begin{matrix} x_1 & y_1 \\ x_2 & y_2 \\ x_3 & y_3 \end{matrix}$$

The computation goes as follows: First, you multiply the three diagonals reading from left to right and add them together. Then you multiply the three diagonals from right to left, add them together, and subtract that total from the first total.

$$Area = x_1 \cdot y_2 \cdot 1 + y_1 \cdot 1 \cdot x_3 + 1 \cdot x_2 \cdot y_3 - \left(1 \cdot y_2 \cdot x_3 + x_1 \cdot 1 \cdot y_3 + y_1 \cdot x_2 \cdot 1 \right)$$

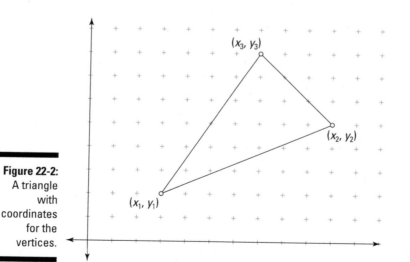

Figure 22-2:
A triangle with coordinates for the vertices.

Chapter 23

Ten Sine Curves Representing Monthly Temperatures

· ·

In This Chapter

▶ Finding the average temperatures in select cities

▶ Predicting temperatures to plan your vacations

· ·

The average monthly temperatures in cities around the world are periodic in nature, meaning that with the season changes, the temperatures change and are relatively predictable. Of course, one year can be much different from another — one summer can be hot and humid in Peoria, Illinois, and the next can be cool and rainy. The functions in this chapter are based on average temperatures taken over about a 30-year period. I used a graphing calculator — with its curve-fitting capability — to create these functions. All you have to do is pick a month (with January being Month 1; February, Month 2; and so on), and you can see the expected temperatures for that month.

To get a feel for what the temperatures will be like, you mainly want to look for the average temperature and the range that the temperatures will fall between. In a trigonometric function with the general format $T(m) = A \sin(Bm - C) + D$, the average yearly temperature is represented by the D value, and the amount the temperatures go above and below that value is represented by A. The m is the variable where you put in the number for the month. The B and C have to do with the change in the period of the sine function and its phase shift — where the curve starts.

Boston, Massachusetts

The temperatures in Boston model after the equation $T(m) = 21 \sin(0.51m - 2.18) + 51$. The average yearly temperature is 51 degrees Fahrenheit, and the temperature fluctuates by 21 degrees — meaning it'll go down to 30 degrees and up to 72 degrees on average. A city along the Atlantic Ocean has relatively moderate temperatures.

Chicago, Illinois

Chicago's temperatures follow the model $T(m) = 26\sin(0.48m - 1.84) + 47$. Its average temperature is lower than that of Boston, but it fluctuates more — by 26 degrees either way of 47 degrees. This place is also called the Windy City, and those gusts will make it seem cooler at times.

Denver, Colorado

If you want the average temperature in Denver, use $T(m) = 21\sin(0.55m - 2.38) + 51$. This formula is almost the same as Boston's — just a bit of a shift. What's the average temperature in Denver in June? Because June is the sixth month, let m be equal to 6 in the formula, and you find that the temperature in Denver averages almost 68 degrees in June. Sounds good to me!

Fairbanks, Alaska

Fairbanks is a fair bit up there! To find the average temperatures, use $T(m) = 38\sin(0.52m - 1.98) + 27$. The average yearly temperature is below freezing, at a cool 27 degrees Fahrenheit. It goes up and down 38 degrees from the midpoint, for average highs of 65 degrees and average lows of –11 degrees.

Galveston, Texas

From Fairbanks, head south — a *long* way. Galveston's temperatures model after the equation $T(m) = 16\sin(0.48m - 1.90) + 68$. This city averages 68 degrees. What's the average temperature in Galveston in February? Put in 2 for the m in the formula, and you get about 55 degrees — this city is warmer than many places but a bit chilly for a winter retreat.

Honolulu, Hawaii

Another warm place is Honolulu. To get the average temperatures, use $T(m) = 4\sin(0.50m - 2.36) + 77$. The average temperature is 77 degrees, and it doesn't change much from that. A 4-degree shift upward and downward makes Honolulu a consistently wonderful place to be.

Miami, Florida

In Miami, the average temperatures follow the model $T(m) = 8\sin(0.47m - 1.87) + 75$. The average temperature is 75, and it goes 8 degrees higher or lower than that. Now, I know that you've been in Miami when the temperature is warmer than 83, but these averages are for a whole month, which tends to even things out. And then you have to factor in a little concept called *humidity*.

San Diego, California

Is any place better than Honolulu? Try San Diego. Its average temperatures follow the equation $T(m) = 7\sin(0.55m - 2.66) + 65$. With an average temperature of 65 and a variation of 7 degrees, you can't go wrong. Low humidity. Little smog. This place is heaven on earth.

San Juan, Puerto Rico

Surrounded by ocean, San Juan is tucked away on the beautiful island of Puerto Rico. You can find its temperatures with the equation $T(m) = 3\sin(0.49m - 2.22) + 80$. They don't vary much — by just 3 degrees. If you're looking for big changes in the seasons, San Juan isn't the place to be.

Sault Ste. Marie, Michigan

Here's a city close to Canada and a very cold lake. Its temperatures follow the model $T(m) = 26\sin(0.49m - 2.00) + 38$. The average temperature is above freezing — but just barely. The temperatures vary by 26 degrees above and below. This city is a great place to be in the summer. In August, the average temperatures are between 62 and 63 degrees. The nights are cool for sleeping.

Appendix

Trig Functions Table

● ●

*Y*ou can use this table of values for trig functions when solving problems, sketching graphs, or doing any number of computations involving trig. I refer to this table in many of the chapters throughout this book. The values here are all rounded to three decimal places. If you need values for angles larger than 90 degrees, check out Chapter 8 to see how to use reference angles. If you need radians instead of degrees, Chapter 5 has information on changing from radians to degrees and back again.

θ	sin θ	cos θ	tan θ	cot θ	sec θ	csc θ
0°	.000	1.000	.000	Undefined	1.000	Undefined
1°	.017	1.000	.017	57.290	1.000	57.299
2°	.035	.999	.035	28.636	1.001	28.654
3°	.052	.999	.052	19.081	1.001	19.107
4°	.070	.998	.070	14.301	1.002	14.336
5°	.087	.996	.087	11.430	1.004	11.474
6°	.105	.995	.105	9.514	1.006	9.567
7°	.122	.993	.123	8.144	1.008	8.206
8°	.139	.990	.141	7.115	1.010	7.185
9°	.156	.988	.158	6.314	1.012	6.392
10°	.174	.985	.176	5.671	1.015	5.759
11°	.191	.982	.194	5.145	1.019	5.241
12°	.208	.978	.213	4.705	1.022	4.810
13°	.225	.974	.231	4.331	1.026	4.445
14°	.242	.970	.249	4.011	1.031	4.134
15°	.259	.966	.268	3.732	1.035	3.864

(continued)

θ	$\sin \theta$	$\cos \theta$	$\tan \theta$	$\cot \theta$	$\sec \theta$	$\csc \theta$
16°	.276	.961	.287	3.487	1.040	3.628
17°	.292	.956	.306	3.271	1.046	3.420
18°	.309	.951	.325	3.078	1.051	3.236
19°	.326	.946	.344	2.904	1.058	3.072
20°	.342	.940	.364	2.747	1.064	2.924
21°	.358	.934	.384	2.605	1.071	2.790
22°	.375	.927	.404	2.475	1.079	2.669
23°	.391	.921	.424	2.356	1.086	2.559
24°	.407	.914	.445	2.246	1.095	2.459
25°	.423	.906	.466	2.145	1.103	2.366
26°	.438	.899	.488	2.050	1.113	2.281
27°	.454	.891	.510	1.963	1.122	2.203
28°	.469	.883	.532	1.881	1.133	2.130
29°	.485	.875	.554	1.804	1.143	2.063
30°	.500	.866	.577	1.732	1.155	2.000
31°	.515	.857	.601	1.664	1.167	1.972
32°	.530	.848	.625	1.600	1.179	1.887
33°	.545	.839	.649	1.540	1.192	1.836
34°	.559	.829	.675	1.483	1.206	1.788
35°	.574	.819	.700	1.428	1.221	1.743
36°	.588	.809	.727	1.376	1.236	1.701
37°	.602	.799	.754	1.327	1.252	1.662
38°	.616	.788	.781	1.280	1.269	1.624
39°	.629	.777	.810	1.235	1.287	1.589
40°	.643	.766	.839	1.192	1.305	1.556
41°	.656	.755	.869	1.150	1.325	1.524

θ	$\sin \theta$	$\cos \theta$	$\tan \theta$	$\cot \theta$	$\sec \theta$	$\csc \theta$
42°	.669	.743	.900	1.111	1.346	1.494
43°	.682	.731	.933	1.072	1.367	1.466
44°	.695	.719	.966	1.036	1.390	1.440
45°	.707	.707	1.000	1.000	1.414	1.414
46°	.719	.695	1.036	.966	1.440	1.390
47°	.731	.682	1.072	.933	1.466	1.367
48°	.743	.669	1.111	.900	1.494	1.346
49°	.755	.656	1.150	.869	1.524	1.325
50°	.766	.643	1.192	.839	1.556	1.305
51°	.777	.629	1.235	.810	1.589	1.287
52°	.788	.616	1.280	.781	1.624	1.269
53°	.799	.602	1.327	.754	1.662	1.252
54°	.809	.588	1.376	.727	1.701	1.236
55°	.819	.574	1.428	.700	1.743	1.221
56°	.829	.559	1.483	.675	1.788	1.206
57°	.839	.545	1.540	.649	1.836	1.192
58°	.848	.530	1.600	.625	1.887	1.179
59°	.857	.515	1.664	.601	1.972	1.167
60°	.866	.500	1.732	.577	2.000	1.155
61°	.875	.485	1.804	.554	2.063	1.143
62°	.883	.469	1.881	.532	2.130	1.133
63°	.891	.454	1.963	.510	2.203	1.122
64°	.899	.438	2.050	.488	2.281	1.113
65°	.906	.423	2.145	.466	2.366	1.103
66°	.914	.407	2.246	.445	2.459	1.095
67°	.921	.391	2.356	.424	2.559	1.086

(continued)

θ	$sin\,\theta$	$cos\,\theta$	$tan\,\theta$	$cot\,\theta$	$sec\,\theta$	$csc\,\theta$
68°	.927	.375	2.475	.404	2.669	1.079
69°	.934	.358	2.605	.384	2.790	1.071
70°	.940	.342	2.747	.364	2.924	1.064
71°	.946	.326	2.904	.344	3.072	1.058
72°	.951	.309	3.078	.325	3.236	1.051
73°	.956	.292	3.271	.306	3.420	1.046
74°	.961	.276	3.487	.287	3.628	1.040
75°	.966	.259	3.732	.268	3.864	1.035
76°	.970	.242	4.011	.249	4.134	1.031
77°	.974	.225	4.331	.231	4.445	1.026
78°	.978	.208	4.705	.213	4.810	1.022
79°	.982	.191	5.145	.194	5.241	1.019
80°	.985	.174	5.671	.176	5.759	1.015
81°	.988	.156	6.314	.158	6.392	1.012
82°	.990	.139	7.115	.141	7.185	1.010
83°	.993	.122	8.144	.123	8.206	1.008
84°	.995	.105	9.514	.105	9.567	1.006
85°	.996	.087	11.430	.087	11.474	1.004
86°	.998	.070	14.301	.070	14.336	1.002
87°	.999	.052	19.081	.052	19.107	1.001
88°	.999	.035	28.636	.035	28.654	1.001
89°	1.000	.017	57.290	.017	57.299	1.000
90°	1.000	.000	Undefined	.000	Undefined	1.000

Index

● *D* ●